New Frontiers in Colloid Science
A Celebration of the Career of Brian Vincent

New Frontiers in Colloid Science
A Celebration of the Career of Brian Vincent

Edited by

Simon Biggs
School of Process, Environmental & Materials Engineering,
University of Leeds, Leeds, UK

Terence Cosgrove
School of Chemistry, University of Bristol, Bristol, UK

Peter Dowding
Infineum UK Limited, Abingdon, Oxfordshire, UK

RSCPublishing

The proceedings of New Frontiers in Colloid Science: a conference to celebrate the career of Professor Brian Vincent held on 1–2 April 2008 at Lord's Cricket Ground, London, UK

Special Publication No. 314

ISBN: 978-0-85404-113-8

A catalogue record for this book is available from the British Library

Published by The Royal Society of Chemistry,
Thomas Graham House, Science Park, Milton Road,
Cambridge CB4 0WF, UK

Registered Charity Number 207890

For further information see our web site at www.rsc.org

Preface

Brian Vincent retired from the position of Leverhulme Professor of Physical Chemistry in the University of Bristol on 31 December 2007. This volume is a collection of articles that were presented at his retirement conference held in April 2008 at Lord's Cricket Ground. Brian's career began in Bristol as a chemistry undergraduate in 1961. After obtaining a first class honours degree in chemistry and an MSc in surface chemistry and colloids with commendation, he began a PhD under the supervision of Ron Ottewill in 1965. After successfully defending his thesis, Brian then moved to Wageningen to work in Hans Lyklema's laboratory on a Royal Society Fellowship. This was to be the start of a lifelong friendship and scientific collaboration with the Colloid and Physical Chemistry Group there. After returning to the UK in 1969 Brian took up a joint post at ICI Paints Division and Bristol University which in 1972 became a full-time lecturing position at Bristol. In 1982 he obtained a DSc in chemistry and was promoted to reader in 1983. In 1992 he was appointed to a chair in physical chemistry and in 1993 became the fifth Leverhulme Professor in Physical Chemistry. In 1994, together with Dr Jim Goodwin, Brian founded the very successful Bristol Colloid Centre, an organisation which carries out short-term research and consultancy work for industry.

Brian served as chair of the SCI Surface Chemistry and Colloids Group from 1988 to 1992 and has also been a member of the SCI Council and Executive Committee. He was elected chair of the RSC Colloid and Interface Science Group in 2000 and has served on the Council of the Faraday Division of the RSC. He served as president of IACIS from 2003 to 2006. Brian has played a prominent role in bringing together the various societies that influence and drive colloid science throughout the world. Another prominent activity which Brian initiated are the European student conferences in colloid science which began in 1981 as a joint activity between Bristol and Wageningen but now involve many of the leading colloid groups in Europe.

Brian has received many honours including the SCI Founder's Lecture and Award, the SCI Distinguished Service Award, the RSC Award in Surface and

New Frontiers in Colloid Science: A Celebration of the Career of Brian Vincent
Edited by Simon Biggs, Terence Cosgrove and Peter Dowding
© The Royal Society of Chemistry 2008

Colloid Chemistry and the Rehbinder Lecture and Medal (Moscow). He has had numerous invitations to give lectures throughout the world and has published over 250 papers, articles, books and patents during his career. He has also been very active at the academic/industry interface and has been a consultant for many of the leading companies that use colloid technology worldwide.

The contributors to this symposium volume comprise a selection of Brian's past students and postdoctoral researchers who have themselves pursued academic careers and other colleagues with whom he has worked extensively. The range of the subject material highlights Brian's own very broad interests in colloid science; it also reflects his long-standing interest in both the academic fundamentals as well as practical applications of the subject. This volume is dedicated to Brian in recognition of his considerable contribution to the world of colloid science and to the guidance and inspiration he has given to many future generations of colloid scientists. We are certain that in his retirement Brian will continue to pursue the subject that he has made his own.

Terence Cosgrove, Simon Biggs, Peter Dowding

Contents

New Frontiers in Colloid Science: A Celebration of the Career of Brian Vincent
Edited by Simon Biggs, Terence Cosgrove and Peter Dowding
© The Royal Society of Chemistry 2008

Chapter 8   Heteroflocculation Studies of Colloidal Poly(*N*-isopropyl-acrylamide) Microgels with Polystyrene Latex Particles: Effect of Particle Size, Temperature and Surface Charge
Martin J. Snowden, Louise H. Gracia and Hani Nur

Chapter 9   Surface Modification, Encapsulation and Coating: A Career Built on Graft
David Fairhurst

Contributors

Steven P. Armes, *Department of Chemistry, Dainton Building, University of Sheffield, Brook Hill, Sheffield S3 7HF, UK*

Simon Biggs, *Institute of Particle Science and Engineering, School of Process, Environmental & Materials Engineering, University of Leeds, Leeds LS2 9JT, UK*

Pierre Bussierre, *Department of Chemical Engineering and Chemical Technology, Imperial College London, Prince Consort Road, London SW7 2AZ, UK*

Peter Dowding, *Infineum UK Limited, Milton Hill Business and Technology Centre, Abingdon OX13 6BB, UK*

David Fairhurst, *Colloid Consultants Ltd, Congers, NY 10920-1834, USA*

Gerard Fleer, *Laboratory of Physical Chemistry and Colloid Science, Wageningen University, 6703 HB Wageningen, The Netherlands*

Louise H. Gracia, *Medway Sciences, University of Greenwich, Central Avenue, Chatham Maritime, Kent ME4 4TB, UK*

Paul Luckham, *Department of Chemical Engineering and Chemical Technology, Imperial College London, Prince Consort Road, London SW7 2AZ, UK*

Paulo Nassari, *Department of Chemical Engineering and Chemical Technology, Imperial College London, Prince Consort Road, London SW7 2AZ, UK*

Hani Nur, *Medway Sciences, University of Greenwich, Central Avenue, Chatham Maritime, Kent ME4 4TB, UK*

Neil Patel, *Department of Chemical Engineering and Chemical Technology, Imperial College London, Prince Consort Road, London SW7 2AZ, UK*

Clive A. Prestidge, *Ian Wark Research Institute, ARC Special Research Centre for Particle and Material Interfaces, University of South Australia, Mawson Lakes, SA 5095, Australia*

John Ralston, *Ian Wark Research Institute, University of South Australia, Mawson Lakes Campus, Mawson Lakes, Adelaide, SA 5095, Australia*

Martin J. Snowden, *Medway Sciences, University of Greenwich, Central Avenue, Chatham Maritime, Kent ME4 4TB, UK*

Carlo Strazza, *Department of Chemical Engineering and Chemical Technology, Imperial College London, Prince Consort Road, London SW7 2AZ, UK*

Brian Vincent, *School of Chemistry, University of Bristol, Bristol BS8 1TS, UK*

Chapter 1

A Journey Through Colloid Science

Brian Vincent

SCHOOL OF CHEMISTRY, UNIVERSITY OF BRISTOL, BRISTOL
BS8 1TS, UK

1.1 Early Days

Colloid science was not part of the chemistry curriculum at Bristol University
during my undergraduate years (1961–4). However, interfacial science was very
strongly established at Bristol, both in teaching and research. The Leverhulme
Chair in Physical Chemistry had been established in Bristol in 1919, in part to
keep J.W. McBain (a Canadian) from being tempted back to North America.
McBain had been appointed as a lecturer in chemistry in the old University
College of Bristol in 1906, three years before the University of Bristol received its
charter and four years before the first, purpose-built university chemistry building
in Woodland Road was completed. McBain rapidly established his name internationally for his work on the association of soap molecules in solution; hence the
approach to Lord Leverhulme, who had built his soap factory (and Port Sunlight
village for his workers) in Cheshire. Eventually, in 1927, McBain did succumb to a
position in the USA, at Stanford University. In his place, W.E. Garner (an expert
in solid-state chemistry and heterogeneous catalysis) was appointed to the
Leverhulme Chair, and he was followed in turn by Douglas Everett in 1954.
Douglas's primary early interests were in gas adsorption (especially the role of
porosity) and in adsorption from solution.

I joined Douglas's very large research group in the academic year 1963–4 to
carry out my final year undergraduate research project, although my project was
supervised on a day-to-day basis by Alan Leadbetter (as was that of a con-
temporary chemistry student and good friend to this day, Terry Blake). My
project was concerned with determining the surface tension, at low temperatures,
of liquid ethane and nitrous oxide for some porosity studies Douglas was doing at
that time. It required the building of a high-vacuum rig, for distilling the two

New Frontiers in Colloid Science: A Celebration of the Career of Brian Vincent
Edited by Simon Biggs, Terence Cosgrove and Peter Dowding
© The Royal Society of Chemistry 2008

liquids into a differential capillary rise cell. This was a challenging task for an undergraduate, but I received splendid help from a young trainee glassblower at that time, one Jim Goodwin, who, much later, after having obtained his PhD, transferred to the academic staff at Bristol and subsequently became an internationally renowned rheologist! Despite only obtaining reliable results pretty close to the end of my allocated project time, that work did result in my first scientific publication.[1] In addition to Terry Blake, there were several others doing their undergraduate projects in physical chemistry at the same time, and with whom I have remained in contact over the years: John Comyn who became a professor in polymer science at De Montfort University, David Billett who worked at Tioxide before becoming a school teacher, Julian Waters who went to ICI Paints and Barry Ingram who had a long career with Proctor and Gamble.

During my final undergraduate year, in early 1964, a new lecturer was appointed in physical chemistry. He initially set up his equipment in our laboratory, where he also located his desk and his cohort of young researchers from Cambridge he had brought with him. That was Ron Ottewill, who subsequently went on to become the fourth Leverhulme Professor of Physical Chemistry, succeeding Douglas in 1982. Ron brought with him to Bristol expertise in colloid science. Douglas had hired him primarily to set up, along with Dr Aitken Couper, a new postgraduate, one-year master of science course in surface chemistry and colloids, by advanced study and research. Despite having applied to do a postgraduate teacher-training course, I was "informed", as was Dave Billett, by Douglas Everett that we were to be "guinea pig" students on that very first MSc course in 1964–5 (the course was to last for more than 30 years!). There were six of us in total in that first year. The first two academic terms (twenty weeks) were spent doing lectures and "set experiments" (although most of these had to be set up as we went along!). The summer was spent doing a four-month research project, and I was allotted to work in Ron Ottewill's group. That was where and when my introduction to the world of colloid science began, and I followed this up working with Ron for my PhD (1965–8).

At the time of his move to Bristol, Ron had already established himself as a leading international expert in the field of polymer latices prepared by emulsion polymerisation. My PhD project was concerned with studying the properties of polystyrene latex particles in alcohol–water media. The work fell in two parts. The first part was concerned with determining the composite adsorption isotherms of a series of alcohol (methanol to *n*-butanol)–water mixtures onto polystyrene particles, together with contact angle (sessile drop) studies of these same liquid mixtures onto thin polystyrene films, prepared by dissolving the latex particles in methylethylketone and evaporating a liquid film of the resulting solution on a glass slide placed in an oven. That work[2] was followed up by me later with a theoretical analysis of the data, which led to a value for the surface tension of polystyrene ($55 \pm 3\,\mathrm{mN\,m^{-1}}$), and which was published as a conference proceedings.[3] However, in hindsight, a much longer-term research interest was to develop from the second part of my PhD project. This was concerned with the coagulation (and corresponding electrophoretic mobility) studies of polystyrene latex particles (with surface carboxylic acid groups) in alcohol–water mixtures.[4] We showed that the critical coagulation concentration (c.c.c.) for $Ba(ClO_4)_2$ passed through a maximum with increasing concentration for each of the lower alcohols, from methanol to *n*-butanol. This was mirrored by corresponding maxima in the (negative) electrophoretic mobility of the particles and also in the relative adsorption of the alcohol molecules concerned on the particles. The explanation offered was that, at low concentrations of each alcohol species, the specific adsorption of the Ba^{2+} ions on the COO^- groups on the particles was reduced by the adsorption of the alcohol molecules, raising the Stern potential, but that at higher alcohol concentrations the subsequent decrease in electrophoretic mobility, and hence the c.c.c. value, was due to a decrease in the ionisation of the COOH groups (to COO^-) as the dielectric constant of the medium was reduced.

After my PhD studies were finished, I was fortunate to be awarded a Royal Society fellowship for one year, and, on the advice of Ron Ottewill, went to work with a young (relatively!) professor in Wageningen University in the Netherlands: Hans Lyklema. I was Hans' second postdoc, Tharwat Tadros being the first. There I was introduced to the world of silver iodide (AgI) dispersions. Such dispersions had been a hot topic of research in both Utrecht (Hans' alma mater) and Wageningen, as they seemed to be excellent model particles, whose charge was controlled by the pAg of the dispersion medium, rather than the pH, as in most other aqueous colloidal systems. AgI is a known ice-nucleator (in seeding clouds to induce rainfall, for example), so it was of interest to study AgI dispersions as the temperature was reduced towards 0 °C. The potentiometric titration data and c.c.c. values, in the presence of a series of alkali metal nitrate salts, suggested that there is indeed an increase in "water structure" at the AgI/water interface as the temperature approaches 0 °C, at least in the region of the zero point of charge of the AgI particles.[5] Measurements on the effect of adding *n*-butanol to aqueous AgI dispersions were also carried out,[6] in a follow-up to my PhD work. Again, a maximum in the c.c.c. for 1 : 1 electrolytes was found with increasing *n*-butanol concentration, with the evidence again pointing towards an initial displacement of specifically adsorbed counter-ions from the Stern layer by the preferentially adsorbed *n*-butanol molecules. That part of the work was carried out in collaboration with a young student, who later became a professor and world expert on proteins at interfaces: Willem Norde.

My year in Wageningen (1968–9) laid the foundations for subsequent collaborations in later years, not only with the Wageningen group (in particular a PhD student of Hans at the time, Gerard Fleer), but also with Tharwat Tadros, whom I came to know very well during that year. In fact, after a year working in the TNO laboratories in Delft, both Tharwat and I joined ICI in 1969, Tharwat at the ICI Plant Protection Division laboratory at Jealotts Hill, near Bracknell (now the Syngenta laboratory), and myself some 10 or so miles away, at the ICI Paints Division laboratory at Slough. In fact initially our two young families each occupied an apartment in an ICI-owned block of apartments in Maidenhead. I stayed at ICI Paints for almost three years, and during that time was privileged to work with some great scientists (and great people!). The praises of industrially based scientists rarely get sung sufficiently, but I learned so much working in the Slough laboratory (principally with "Ossie" Osmond, Fred Waite, Derek Walbridge, Ron Lambourne and Andrew Doroszkowski). In particular, I was introduced to the synthesis, characterisation and stability of polymer particles in non-polar solvents, and the concept of steric stabilisation. Working in a major industrial laboratory at that time was not so different from working in an academic laboratory. The laboratories were often directed by academically minded persons, who encouraged one to work on longer-term, more fundamental projects (with some relevance, of course, to the company's products!), and also to publish!

One project I worked on at ICI concerned extending Marjorie Vold's earlier theoretical analysis[7] for calculating the van der Waals interaction between colloidal particles carrying adsorbed or grafted polymer layers, having a segment density distribution.[8,9] Another project I worked on, which was later to become a major research theme of mine in my early years at Bristol, was the question of why some solvent-based, supposedly gloss paints lose some of their gloss during drying, especially under poor drying conditions. It was known that this was associated with aggregation of the pigment particles within the drying film. To help unravel this problem we worked on some model aqueous systems, polystyrene particles carrying terminally grafted poly(ethylene oxide) chains (PS-*g*-PEO particles).[10] These particles aggregated over a fixed concentration range of PEO + water mixtures. This general phenomenon, *i.e.* the aggregation of colloidal particles in the presence of non-adsorbing polymers, eventually became known as "depletion flocculation". I shall return to this theme in Section 1.3.3. We also tried to understand why the (reversible) aggregation we observed with our aqueous PS–PEO particle dispersions, in the presence of free PEO chains, was more akin to a phase separation process than classical (irreversible) coagulation (see Section 1.3.2).

My position at ICI also involved an appointment as a part-time lecturer at Bristol University. In fact Don Napper had held a similar dual position (ICI + Bristol) before me, prior to returning to Sydney. By the early 1970s, the first major international oil crisis was upon us and we had experienced the three-day working week in the UK, imposed by the Government in response to power shortages brought about by the miners' strike. This brought changes to industrial practice and, in particular, managers decided that research needed to be more "focused"; "academic"-style research from then on would be less and less tolerated in industrial R&D laboratories—world wide. So for me it was time to move on from ICI. Fortunately for me, Frank Stone moving to Bath University from Bristol created a vacancy in the physical chemistry staff. So I applied and was fortunate to be selected, and in 1972 I moved full time to Bristol, as a lecturer.

My research over the last 35 years at Bristol has evolved in many different directions. In many cases new research themes have been initiated by something discussed at a conference or by tackling some challenge proposed by a colleague in industry. Indeed, I would estimate that roughly half my research sponsorship over the years (which has supported about 90 PhD students and about 50 postdocs/visitors) has come from industry. I have always enjoyed the challenge of solving technological challenges, as well as opening up new basic, scientific directions. However, I have always insisted that any research that we did in collaboration with industry should be of a nature that would lead to eventual publication in the open literature. In order to describe the last 35 years' research of my group at Bristol, I will from here on resort to research "themes", rather than to continue to describe the work in chronological order.

1.2 Novel Colloidal Systems

1.2.1 Particle Synthesis

For maybe forty years after the Second World War, one could argue that colloidal *physics* was the most dominant and influential area of colloid science. Not only did an extensive literature develop around the theory of the physical properties of colloidal systems, mainly in the wake of the "DLVO" theory[11,12] of the aggregation of colloidal particles, but there were considerable developments also in the various areas of physics which underpinned the new instrumentation available to experimental colloid scientists. However, during that period the actual systems available for study were rather limited: latex particles (especially PS latex), metal oxides (*e.g.* silica) and inorganic salts (*e.g.* AgI) were the prime candidates for most researchers in the field. Each system had its advantages and disadvantages. Studies of classical liquid–liquid dispersions (emulsions) were also limited in that polydispersity was a real problem, although the much more monodisperse *microemulsions* were being increasingly investigated. In the last twenty to thirty years, colloidal *chemistry* has made a strong comeback. We now have available a large array of particles which have well-defined size and shape, as well as structure, composition and physical properties; monodisperse liquid droplets are also now available. These developments have been stimulated, in part, by the challenges that have been thrown up by so-called "nanotechnology".

I have always tried to maintain a strong synthetic aspect to my research at Bristol, and in the sections that follow I describe some of the main types of colloidal system on which we have worked. There have been a number of systems we have worked on where we have tried to understand, often in collaboration with industrial colleagues, the physical principles underlying the formation of specific types of colloidal particles. One example would be the work that Jenny Saunders did for Laporte Ltd in trying to control the properties of synthetic Laponite clay particles by systematically varying the ratios of chemicals used in the synthesis, and then studying the structure of the particles with small-angle X-ray scattering.[13] Tim Muster, in collaboration with Rhodia, studied particle formation (and subsequent gel formation) on adding oligomeric phthalate-based polyesters to water.[14] David Voisin, sponsored by Unilever, similarly studied particle (and gel) formation in mixtures of cationic polyelectrolytes and anionic surfactants.[15]

Polymer latex particles, as discussed earlier, have proven to be a major class of model systems, widely studied in colloid science. Much of the earlier work on the synthesis of specific types of latex particles in my group will be described in later sections, *e.g.* "hairy" latex particles carrying terminally attached polymer

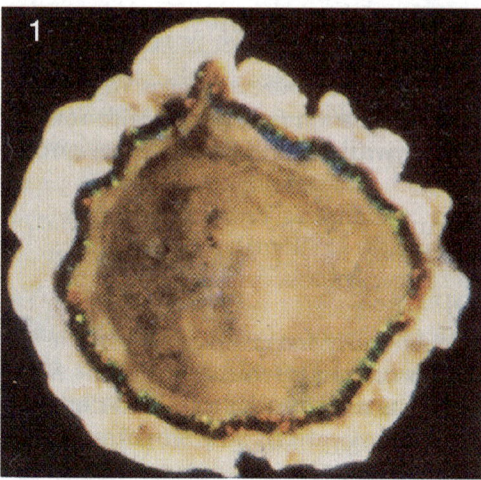

Figure 1.1   Section through a Tertiary (Palaeocene) species of the selaginellalean megaspore *Erlansonisporites*, illustrating the iridescence of the exine layer.[20]

chains (Section 1.2.2), electrically conducting latex particles (Section 1.2.3), very small ($\sim$ 10 nm) polystyrene particles (Section 1.2.4), very large ($\sim$ 100 µm) monodisperse, *porous* polystyrene particles (Section 1.2.5) and swellable latex particles (Section 1.2.7). However, we have returned very recently to a couple of "classical" themes in this area. In one project Ciaran Martin has been trying to advance our understanding, in collaboration with AGFP, of the mechanisms underlying the formation of polytetrafluoroethylene particles by dispersion polymerisation of the dissolved, gaseous monomer.[16] In the second topic Aaron Olsen, from Newcastle University in Australia, has been synthesising amphoteric polystyrene particles, with a controlled iso-electric point, by using a mixed initiator system,[17] and trying again to understand the mechanisms involved, following up some earlier studies by Ron Ottewill.[18]

One fascinating area of particle formation in *biosystems* that I was closely involved with was undertaken by Alan Hemsley, a palaeontologist at Cardiff University.[19–22] The pore walls of certain pollen and spore species exhibit iridescent colours (see Figure 1.1).[20] This is due to the crystallisation of natural latex particles in the walls, subsequent to their formation in aqueous droplets which eventually evaporate. We tried to simulate the particle formation and subsequent crystallisation processes using synthetic materials.

In the following sections areas of research on novel colloidal particles will be described which have been sustained over longer time periods.

1.2.2   Particles with Terminally Grafted Polymer Chains

As will be described in some detail later (Section 1.3.1), one of the major recurring themes in my research over the years has been the study of polymers *physically adsorbed* on surfaces. However, it also seemed of interest, early in my research at Bristol, to initiate studies on polymers which had been *chemically anchored* to a particle surface. My second PhD student, Keith Bridger, had received an EPSRC CASE (cooperative award in science and engineering) to work with myself and David Fairhurst, then at the UK Ministry of Defence laboratories at Porton Down in the UK, but subsequently to become a leading US-based colloid scientist, and still a good friend. We decided to study the grafting of "living" polymer chains (*e.g.* PS) to suitably modified silica surfaces in non-polar solvents, using anionic polymerisation methods. We were able to terminally graft PS chains of (number average) molecular weight in the range 2000 to 50 000 to the silica particles.[23,24] This led to other silica-grafted polymer systems produced by our group, *e.g.* with grafted PEO,[25] and with grafted polydimethylsiloxane (PDMS).[26]

I have already referred, in Section 1.1, to polystyrene particles with terminally attached PEO chains. These were first prepared at ICI, with early reports appearing in the patent literature and only much later in the open literature.[27] The method was modified by Cowell and myself.[28] It involves the dispersion copolymerisation of styrene monomer with a *macro-monomer* (*i.e.* a PEO chain having a terminal vinyl group). Keith Ryan, a postdoc working with Terry Cosgrove and myself, later described a method for post-grafting PEO to *preformed* PS particles.[29]

1.2.3 Electrically Conducting Particles

In order to investigate how the conductivity of colloidal particles affects their properties, *e.g.* their stability to aggregation and their electrophoretic mobility, as well as their possible industrial applications, we undertook a series of studies in my group to make different types of electrically conducting particles. Of course, metal particles could be readily synthesised (such as gold and silver), but these were generally small in size (several nanometres typically), were not very monodisperse and were expensive to prepare as concentrated dispersions. Carbon black was also a well-established system showing electrical conductivity, and was used, for example, as an anti-static additive for bulk polymers, but it is hardly a model system, since the particles are themselves agglomerates of smaller particles, and commercial samples therefore have a wide size and shape distribution. Phil Pendleton, another early PhD student of mine (funded by Xerox in Canada), and now a professor at the University of South Australia, therefore set out to synthesise monodisperse, spherical carbon particles, based on the chemical and/or thermal de-hydrochlorination of poly(vinylidene chloride) latex particles.[30,31] Thermal de-hydrochlorination led to particles which were ~90% carbon, but although they were indeed spherical and reasonably monodisperse, they had a high porosity.

Our attention then turned to particles based on electrically conducting polymers. The prototype conducting polymer at that time was polyacetylene, so we set out in the early 1980s to make polyacetylene latex particles. The method used was the dispersion polymerisation of acetylene, dissolved under pressure in tetrahydrofuran (or other solvents), under nitrogen, using the so-called "Lutinger catalyst" [Co(NO$_3$)$_3$ plus NaBH$_4$],and in the presence of a diblock copolymer steric stabiliser [poly(*t*-butylstyrene-*b*-ethylene oxide)]. In the absence of this stabiliser chains of particles appeared to be produced, but in its presence discrete, but rather polydisperse, particles were produced.[32] Particle conductivities (as measured on pellets of the compressed particles) were rather low, typically $\sim 10^{-5} \Omega^{-1} cm^{-1}$, and, moreover, polyacetylene is rather unstable to atmospheric oxidation. When Steve Armes joined my group we took the polyacetylene work somewhat further, and developed specifically designed diblock copolymers to act as steric stabilisers during the particle synthesis, based on the sequential, anionic polymerisation of acetylene and isoprene monomers.[33] However, Steve showed that polypyrrole (PPy) particles proved to be a much more facile and rewarding route to go down. These could be prepared in an aqueous environment using a variety of oxidants (*e.g.* FeCl$_3$), were monodisperse, spherical and had stable, high conductivities ($> 1 \Omega^{-1} cm^{-1}$).[34–36] Gavin Markham, Tim Obey and myself[37] were later able to show that the Hamaker constant for the PPy latex particles is $(45 \pm 6) \times 10^{-20}$ J, *i.e.* comparable to the values reported for metal particles, and were able to confirm a suggestion made previously by Overbeek[38] that conducting particles should behave in a similar manner to non-conducting particles in terms of their electrophoretic mobility, since the influence of the conductance of the particles will effectively be neutralised by the charge polarisation occurring at the surface. In addition to PPy particles we also prepared polyaniline (PAn) particles. These were frequently rod-shaped particles, and as such Liz Cooper studied their possible incorporation into polymer films on surfaces, to impart electrical conductivity to the films, in work funded by ICI.[39,40] However, Joanne Waterson succeeded in making spherical PAn particles.[41] Our own final paper on conducting particles appeared in 1997. This described Gavin Markham's work on the electro-rheological properties of concentrated dispersions of PPy particles in non-polar solvents.[42] With Steve Armes' appointment at Sussex (and now at Sheffield), work on electrically conducting particles largely shifted to that location, and the story continues in the chapter by Steve in this book (Chapter 3).

1.2.4 Microemulsions Based on Block or Graft Copolymers

Microemulsion systems have traditionally been based on surfactants (plus in some cases co-surfactants). However, these are not very robust and may change their structure, *e.g.* on dilution or concentration, or on changing the temperature. It was considered that microemulsions based on AB diblock copolymers might be more robust in this regard. With the expertise developed in anionic polymerisation methods in our group, Martyn Barker set out to investigate the properties of anionically polymerised PS–PEO diblock copolymers[43] in toluene + propan-2-ol + water mixtures. We showed, for example, that stable water-in-oil microemulsions were formed in certain regions of the phase diagram at 20 °C.[44]

Subsequently, in the early 1990s, we returned to this theme, and Zoltan Kiraly (from Szeged University) prepared PDMS–PEO block copolymers by anionic polymerisation.[45] Then Simon Biggs prepared a series of poly (styrene-*b*-vinylpyridine-1-oxide) and poly(dimethylsiloxane-*b*-vinylpyridine-1-oxide) block copolymers, using anionic polymerisation techniques.[46] He subsequently studied their micellar[47] and microemulsion[48] properties. Somehow Simon's research interests went from organic chemistry in Bristol to chemical engineering at Leeds!

Finally under this topic, I should mention the more recent work of Adrian Horgan, who, in studies funded by ICI Paints, prepared graft copolymers based on the free radical copolymerisation of methyl methacrylate or *n*-butyl methacrylate with vinyl-capped PEO monomers.[49] He subsequently used these as templates to make styrene-in-water microemulsions, which could then be polymerised to give very small, sterically stabilised PS nanoparticles (< 10 nm in diameter).[50]

1.2.5 Monodisperse Liquid Droplets

As mentioned earlier, one of the "holy grails" in colloid science has been the quest to produce monodisperse emulsion droplets. Various methods have been suggested over the years, and these are well summarised in the book edited by Bernie Binks on *Modern Aspects of Emulsions Science*.[51] Our own efforts in this regard have fallen into two categories: comminution methods and nucleation and growth methods.

The initial comminution method that Rudy Hengelmolen and I tried was the use of an electro-spray device to form fine spays of droplets of a non-conducting liquid which were collected in an aqueous solution containing a surfactant stabiliser to form oil–water emulsions.[52,53] The droplets were reasonably monodisperse, especially in the spray *per se*, but some droplet break-up tended to occur on their entry into the water phase. Also high droplet concentrations took a very long time to produce by this method. Subsequently, Peter Dowding in my group used a bespoke-built, cross-flow membrane device to make large liquid (in particular, monomer) droplets ($\sim 100 \, \mu m$ or so in size), again reasonably monodisperse, which, if desired, could be fed into a tubular, continuous reactor to make large beads by a suspension polymerisation route. By incorporation of a porogen into the monomer, beads of controlled porosity could be produced.[54–57]

Much more successful in producing monodisperse droplets in the micrometre size range is the method Tim Obey first developed in my group based on nucleation and growth, whereby PDMS–water droplets are prepared *in situ* by the base-catalysed hydrolysis of dimethyldiethoxysilane.[58–60] This reaction is the analogue of the Stöber process for making monodisperse silica particles,[61] in which the tetra-functional silane is used, rather than the di-functional monomer, as in the PDMS case. As with solid polymer latex particles, the PDMS droplets are charge-stabilised,[60] and do not require the addition of surfactant *per se* to stabilise them. This property makes them excellent model emulsion systems, with droplet sizes from a few hundred nanometres to $\sim 10 \, \mu m$ readily prepared. However, if surfactant is used in their synthesis then the equivalent PDMS microemulsion droplets (< 100 nm) may be prepared.[59] By combining the di-functional silane monomer with increasing mole ratios (from 0 to 1) of either the tri- or tetra-functional silane, Mike Goller[62] prepared a series of "particles" which ranged from low-viscosity liquids, through (viscoelastic) microgels (see Section 1.2.7) to silica-like hard particles.

A variation on the use of a chemical reaction to initiate nucleation and growth is to use a physical process, whereby a phase boundary from a one-phase to a two-phase region is crossed. Systems of this type were prepared by Zoltan Kiraly in our group, for making emulsions of low molecular weight PDMS in low molecular weight PEO (or vice versa).[63] At low temperatures mixtures of these two polymers form two phases, but an upper consolute phase boundary may be reached on raising the temperature sufficiently, beyond which the system forms one phase. So, by starting with a system in this one-phase region, and cooling below the upper consolute temperature, nucleation and growth occur, with stable droplets of one phase forming in the second phase. Alternatively, a common solvent for PEO and PDMS, such as ethyl acetate, may be used to form the one-phase region at room temperature, and then evaporated to cross the two-phase boundary line. Using either method, the formation of stable droplets required the presence of a PDMS–PEO diblock copolymer, which Zoltan prepared using sequential anionic polymerisation.[64] The copolymer molecules adsorbed at the emerging liquid–liquid interface. The relative block lengths of the PDMS–PEO copolymer used determined the nature of the droplets formed. Reasonably monodisperse emulsion droplets, typically ~ 1 μm in size could be produced in this way.

1.2.6   Liquid Core–Solid Shell Particles

Encapsulation of liquids containing "active" molecules is widely used in a large range of technologies (including the pharmaceutical, agrochemical, household product and food industries), either to protect the "active" molecules during some part of the process or to effect its triggered release at some later point in the process (*e.g.* a drug, an enzyme, a flavour or perfume molecule, *etc.*). A common triggering mechanism is simple mechanical pressure, to break the shell wall. In that case the rheological characteristics (*e.g.* softness and yield pressure) of the shell, as well as its porosity need to be carefully controlled. We have investigated two classes of shell material, for encapsulating both oil-based and aqueous-based cores: solid inorganic shells and polymer-based shells.

Mike Goller[65] prepared the first encapsulated liquid core–solid shell particles in our group; these were based on the monodisperse PDMS droplets and microgel particles described in the previous section, around which a solid silica shell of controlled thickness could be grown. More recently Michael O'Sullivan,[66] supported by a grant from Schlumberger, developed this method to form silica shells of controlled thickness around PDMS droplets by a simpler procedure, based on mixed silane chemistry; an example of such a PDMS core–silica shell system is illustrated in Figure 1.2.

A good correlation was found between the yield stress of the capsules (as measured by a micromanipulator device, connected to a pressure transducer) and the shell thickness.[66] Michael has also been able to encapsulate water droplets

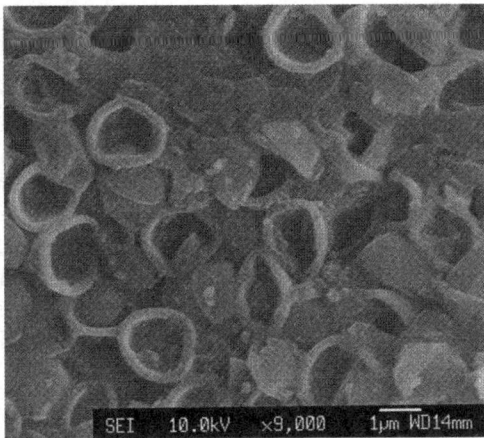

Figure 1.2   SEM image through (sliced) core–shell particles having oil cores (evaporated) and silica shells.[66]

with silica shells using similar silane chemistry.[66] Release of any active species present in the core liquid may be triggered by breaking the inorganic shell.

Replacing the silica shell with a polymer shell offers several potential advantages: (i) a greater variety of methods of preparation; (ii) reduced, potential diffusion of small inorganic molecules (*e.g.* water) or ions through the shell, since silica is known to be microporous; (iii) solution and diffusion of low molecular weight organic "active" species, into and across the shell, to effect their controlled (rather than triggered) release; and (iii) the possibility of effecting dissolution of the polymer shell, or solvent- or temperature-induced swelling for cross-linked polymer shells, as an alternative triggered release mechanism for larger "active" molecules. The challenge is to produce hole-free polymer shells. Andrew Loxley[67] developed a method in our group for producing polymer shells, of controlled thickness and low porosity, around oil cores, based on the *internal* phase separation of polymer from within the oil droplets dispersed in water. The concept is to dissolve the polymer concerned, *e.g.* poly (methyl methacrylate) (PMMA), in a mixture of a poor, non-volatile solvent (*e.g.* hexadecane) and a good, volatile solvent (*e.g.* dichloromethane). Sufficient of the latter solvent is added to just take the polymer into solution. The oil phase is then emulsified into an aqueous phase containing a suitable (but carefully selected) surfactant stabiliser. Evaporation of the volatile solvent is then commenced, effecting phase separation of the polymer. If the surfactant chosen is indeed suitable, then the polymer will collect at the oil–water interface; if not, "acorns" are formed, whereby the polymer separates as a "lump" outside the oil phase, protruding into the aqueous phase. "Suitability" of the surfactant requires that the oil–water interfacial tension is not reduced significantly, compared to the polymer–oil and polymer–water interfacial tensions.[67] Peter Dowding (supported by a grant from Zeneca Agrochemicals, now Syngenta) and Rob Atkin (supported by a grant from P&G, and from Newcastle University in Australia) later developed this method further, and carried out some controlled release studies from various oil core–polymer shell systems of model "active" molecules, such as 4-nitroanisole, which is a good model for organic "active" molecules having a limited solubility in water.[68,69] Here the release mechanism is slow escape by diffusion across the polymer shell. Clearly, if the shell is swellable or dissolvable, then this introduces an additional release mechanism. Manolo Romero-Cano,[70] from Almeria University, studied the controlled release of 4-nitroanisole from oil core–poly(lactic acid) shell particles; the poly(lactic acid) particles dissolve slowly in aqueous media, but this accelerated in acid or base conditions. The Chalmers group in Sweden (Helena Wassenius and Magnus Nydén) later used the oil core–polymer shell particles for NMR diffusion studies of the translational properties of the oil molecules in the cores of core–shell particles.[71]

Rob Atkin developed[72] analogous systems with *aqueous* cores and polymer shells, dispersed initially in an oil phase, but these particles could be subsequently transferred across to a water-based continuous phase using a centrifugation procedure. One then has water droplets, surrounded by a polymer shell, dispersed in water, which would be eminently suitable for the protection and triggered release of, say, enzymes or other proteins into an aqueous environment.

1.2.7 Microgel Particles

Microgel particles are simply cross-linked polymer particles, which show reversible swelling/deswelling behaviour depending on the local thermodynamic conditions, *e.g.* temperature or solvency of the polymer chains, as expressed through the Flory χ-parameter. The mechanism of swelling is essentially an osmotic one, with swelling being opposed by the elastic (entropic) term whose magnitude is controlled by the density of cross-links. If the particles contain charge groups (from ionisable monomers) then ionic strength and also pH (for weak acid and base groups) play an additional role in controlling the swelling/deswelling behaviour (for reviews on microgel particles see refs. 73–75).

The prototype, aqueous-based microgel system is undoubtedly that based on neutral, poly(*N*-isopropylacrylamide) (PNIPAM), first described by Bob Pelton,[76] in which bisacrylamide is used as the cross-linking monomer. High molecular weight PNIPAM has a lower critical solution temperature (LCST) of around 35 °C in water, rendering PNIPAM microgels temperature sensitive. Note that, although PNIPAM is a neutral polymer *per se*, and there is no *bulk* charge within the microgel particles, they do usually carry a peripheral (*surface*) charge resulting from the initiator fragments used to effect the dispersion polymerisation of the particles (usually at temperatures several tens of degrees above the LCST, *i.e.* in the deswollen state).

My own initial interest in microgel systems dates back to the late 1970s when we worked with cross-linked PS particles in ethyl benzene as model systems for studying the depletion flocculation induced on adding (free) PS chains[77] (see Section 1.3.3). My interest was then re-stimulated, some ten years later, when I was approached by BP to discuss possible systems that might be used for "gelling" water pumped down oil wells to displace oil (it is important that the aqueous phase has a similar "viscosity" to the oil, so that "fingering" can be reduced). PNIPAM microgel particles seemed like a good candidate, as similar aqueous-based microgel particles were already being used as rheology modifiers in aqueous paint systems. With funding from BP, I was able to support my first co-worker in the field of microgels, Martin Snowden, now a professor at Greenwich University, where he has successfully developed the use of microgel particles as drug delivery systems. A patent with BP was taken out on the oil well application.[78]

The large body of subsequent work on microgels carried out by my group in Bristol over the last 15 years or so can be divided into two main sections: (i) fundamental studies and (ii) controlled uptake and release. With regard to the former, one of the first areas we studied, in connection with the BP work, was the flocculation of microgel particles in the presence of electrolytes and polyelectrolytes,[79-81] but discussion of this topic is deferred to Section 1.3.2 and 13.3. Microgel particles in the size region of a few hundred nanometres are most readily prepared by dispersion polymerisation methods,[73] but larger particles ("minigels") may be prepared by an inverse-emulsion route, as shown by Peter Dowding.[82]

The basic temperature response of PNIPAM microgel particles is illustrated in Figure 1.3.[83] This shows (Figure 1.3a) how the hydrodynamic diameter (d) responds to changes in temperature (T), for different cross-link densities. The volume phase transition temperature (VPTT) is taken to be the temperature corresponding to the maximum slope, dd/dT; this depends slightly on the cross-link density, but is in the range 32–35 °C. Also shown (in Figure 1.3b) are the spin–spin relaxation times (T_2) for water molecules in the system; the lower T_2 the higher the fraction of "restricted" water, compared to free water molecules. Attempts at determining the density of cross-links within the microgel particles were made by Helen Crowther, in collaboration with Terry Cosgrove's group;[84,85] it was concluded that there is certainly not a uniform distribution of cross-links, and there may well be "hot-spots" consisting of higher cross-link density regions. The solvency response of PNIPAM microgel particles in water + n-alkanol mixtures was also studied by Helen Crowther;[86] she showed that the hydrodynamic diameter passes through a minimum with increasing n-alkanol concentration, due to a co-non-solvency effect. Similar effects were also found[87] for microgel particles based on copolymers of methyl methacrylate and methacrylic acid. The inverse temperature effect, *i.e.* microgels which swell on *heating* rather than cooling, was demonstrated by Philippe Bouillot in our group,[88] with microgel particles based on interpenetrating networks of polyacrylamide (PAM) and poly(acrylic acid) (PAAc). The electrophoretic mobility of PNIPAM microgel particles results from the *surface* charge groups from the initiator used; the values obtained, for example by Mikael Rasmusson (from Chalmers University) in our group,[89] are much harder to interpret than for traditional hard latex particles, because of solvent and ion flow within the particles.

Bulk charge groups in microgels may be introduced in a variety of ways. One way is to use a monomer which itself can ionise with a change in pH, *e.g.* vinylpyridine (VP), which protonates at pH values less than ~4 to give a cationic microgel, as demonstrated by Andrew Loxley,[90] or to copolymerise a

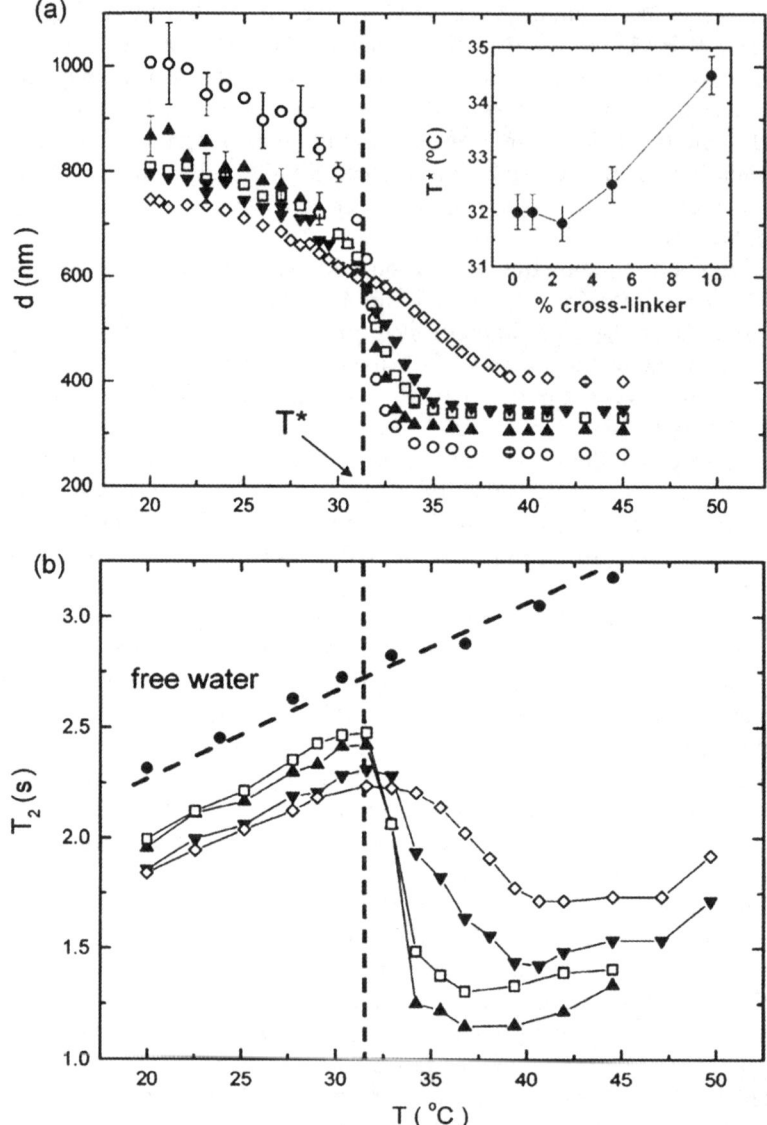

Figure 1.3 PNIPAM microgel particles having different cross-linking monomer
(bisacrylamide) contents, from 0.5 to 10 wt%: (a) hydrodynamic diameter
(*d*) as a function of temperature; (b) spin–spin relaxation time (T_2) for
water molecules as a function of temperature.[83]

polar, ionisable monomer with NIPAM, *e.g.* acrylic acid (AAc), to give an
anionic microgel, as shown, for example, by Gayle Morris, from South
Australia.[91] One may also make amphoteric microgel particles. If mixtures of
strong acidic and basic monomers are used, with equal molar quantities of each
monomer incorporated, the resulting microgel particles are collapsed in water

due to strong electrostatic attraction forces within the particles, but they expand on adding electrolyte (*i.e.* show the reverse electrolyte effect). This effect was demonstrated by Sylvia Neyret (from Strasbourg) in our group for microgel particles based on sodium 2-acrylamide-2-methylpropane sulfonate and (2-(methacryloyloxy)ethyl)ammonium chloride.[92] In contrast, if mixtures of weak acid and weak base groups are used, as in Melanie Bradley's work,[93] then the particles respond to pH changes, with a minimum in size around the iso-electric pH.

The bulk charge magnitude–degree of swelling relationship has been analysed by us in collaboration with the group at Almeria University,[94,95] as has the theory for the electrophoretic mobility of microgel particles with a bulk charge.[96] Andrew Loxley[90] studied the rate of swelling and deswelling of polyvinylpyridine (PVP) microgel particles on changing the pH, using a stopped-flow kinetics apparatus, and subsequently Mel Bradley (from Melbourne University) and Jose Ramos (from San Sebastian)[97] did similar experiments with poly(NIPAM-*co*-AAc) microgel particles. It was shown that typical relaxation times are in the millisecond to second range.

Two more recent developments in microgel systems that have been studied in our group are core–shell particles and microgel monolayers on flat surfaces. The latter topic is discussed further in Section 1.3.5. With regard to the former topic, we have prepared a variety of systems, such as: (i) hard (silica) core–microgel (PVP) shell particles, where the shell responds to pH changes;[98] (ii) cross-linked PMMA core–PEO shell particles, in which the core now swells in water–dioxan mixtures (in collaboration with Sam Kaneda from Shiseido, Japan[99]); and (iii) poly (NIPAM-*co*-VP) core–poly(NIPAM-*co*-dimethylaminoethyl methacrylate [DMAEM]) shells, where the two basic monomers have rather different pK_a values (VP, 4.9; DMAEM, 8.4).[100]

Over the years, our group has studied the uptake and release of a large range of different species, into and out of microgel particles. For example, Gayle Morris and Martin Snowden[101–103] investigated the uptake and release of heavy metal cations from aqueous solutions, using poly(NIPAM-*co*-AAc) particles. The ions were absorbed at high pH and then released again at low pH, in response to changes in the degree of ionisation of the carboxylic acid groups in the AAc moieties. Using a dialysis apparatus, it was possible in this way, to selectively remove toxic ions (such as Pb^{2+} and Cd^{2+}) from water, but leave behind the more "friendly" ions (such as K^+, Na^+, Mg^{2+}, Ca^{2+}).

Although Jane Clarke[77] was actually the first person in my group to investigate the effects of adding free polymer (PS) in solution to microgel particle (cross-liked PS) dispersions (in ethyl benzene), Brian Saunders, now at Manchester University, was the first to actually monitor the resultant particle size changes in the same system.[104] He also studied PNIPAM microgel particles in water to which PEO was added.[105,106]. In all cases he found that the microgel particles deswelled in the presence of the free polymer molecules; this was attributed to the reverse osmotic effect of the polymer chains in solution, which were assumed not to enter the microgel particles. Later, more systematic studies by Mel Bradley and Jose Ramos[97] on the PNIPAM and poly(NIPAM-*co*-AAc) plus PEO systems, particularly at low PEO concentrations, showed that low molecular weight PEO did enter the microgel particles; this was accompanied by *swelling* of the microgel particles, *i.e.* greater swelling than in water alone at 20 °C. The driving force for the absorption is hydrogen bond formation between the ether oxygen atoms in the PEO chains and either the hydrogen on the COOH groups (at low pH) or the hydrogen on the secondary amine units in the NIPAM. With increasing polymer concentration, once the microgel particles had become saturated with PEO, then they underwent the *deswelling*, previously observed by Brian Saunders, as the concentration of free PEO in solution built up. The cut-off molecular weight of PEO chains which could initially enter the microgel particles is linked to their cross-link density. The interpretation of the swelling/deswelling data was corroborated by direct measurements of the absorbed amount of PEO. Some results illustrating the effect of PEO on the microgel particle size are shown in Figure 1.4. The swelling ratio is shown for different molecular weight PEOs added to poly(NIPAM-*co*-AAc) microgel particles. Later, in collaboration with Alex Routh at Sheffield (now at Cambridge) and Alberto Fernandez-Nieves at Harvard, we developed a thermodynamic model of the swelling behaviour.[107]

Our group, and in particular Melanie Bradley, has also investigated extensively the absorption and desorption of surfactant molecules into and out of microgel particles. These studies include non-ionic surfactants,[108] anionic surfactants[109,110] and cationic surfactants.[111] In these cases various mechanisms for absorption have been postulated, with contributions, depending on the system, from hydrogen bonding, hydrophobic bonding and electrostatic interactions. In one case, in collaboration with Julian Eastoe's group, a photo-degradable anionic surfactant was used in an attempt to develop a light-triggered release microgel system.[109]

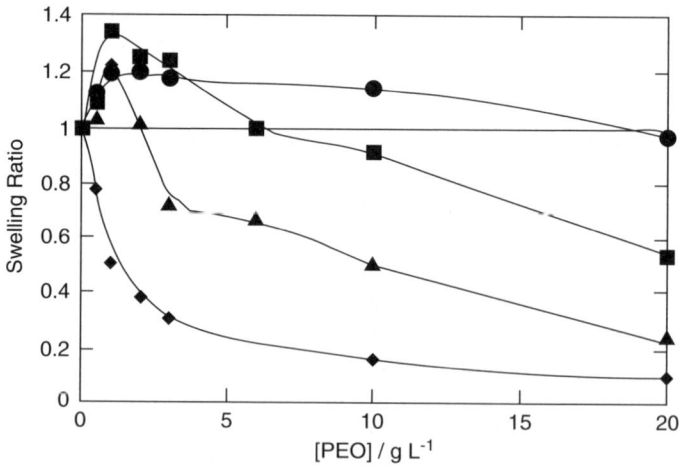

Figure 1.4 Swelling ratio as a function of added PEO concentration, for different molecular weight PEO added to poly(NIPAM-*co*-AAc) microgel particles(10 wt% AAc and 10 wt% cross-linking monomer) at pH = 3: ●, 2000; ■, 20 000; ◆, 100 000; ▲, 300 000.[97]

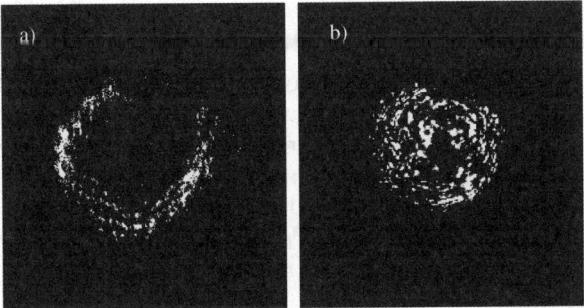

Figure 1.5 Confocal microscope images of CdSe (quantum dot) particles absorbed into cross-linked PS particles: (a) in propanol (a poor solvent for PS); (b) in 60 vol.% chloroform and 40 vol.% propanol (a good solvent mixture). This illustrates that the CdSe particles can penetrate further when the PS particles are more swollen in the better solvent.[112]

More recently we have been investigating the uptake of nanoparticles into (larger) microgel particles. Melanie Bradley[112] looked at quantum dot (cadmium selenide) particles absorbing into large cross-linked PS particles, and followed the penetration of the nanoparticles using confocal microscopy; the degree of penetration directly correlated with the degree of swelling of the PS microgel particles, which could be adjusted by changing the solvent, as illustrated in Figure 1.5. Paul Davies has recently been investigating the uptake of gold nanoparticles into PVP microgel particles.[113]

In all the uptake and release studies we have worked on as regards microgel particles, there are two important considerations. The first is that there must be an attractive interaction for whatever species is under consideration to enter the particle, and, if release is to be triggered, then some change in the local conditions (*e.g.* temperature, solvency, pH, ionic strength, light intensity) has to weaken that attraction. The second consideration only applies to species (*e.g.* polymers or nanoparticles) that are of comparable dimensions to the average "mesh size" within the particles; it is possible, by switching from the swollen to the deswollen state of the microgel particles, to physically *trap* the active species within the microgel particles, preventing their release until the microgel particles are re-swollen.

1.3 Particle Interactions

1.3.1 Electrostatic Interactions

As I have already described in Section 1.1, my own PhD and postdoctoral work was concerned with the stability of charged particles. My interest in electrostatic interactions has continued since then. In particular, the role of charge in non-aqueous systems has been a continuing theme. One of the classical techniques for assessing particle charge is through the measurement of the electrophoretic

mobility of the particles concerned. For aqueous dispersions this is straightforward. For dispersions in non-polar media it is not. When asked by Glaxo
(now GSK) to investigate the possible role of particle charge in stabilising drug
dispersions in propellant fluids (used in asthma sprays for example),[114] I was
convinced that none of the (then) commercially available equipment measured
the particle mobility correctly in non-polar media. It was a chance meeting with
(the late) Klaus Schätzel, then at Keil University, that inspired the development
of the phase analysis light scattering (PALS) instrument by John Miller and
myself at Bristol.[115] Several instrument makers have now incorporated this facility into their equipment. Basically, as the particle charge diminishes, particularly for small particles and/or moderate electrolyte concentrations, particle
diffusion dominates any electrophoretic movement of the particles (unless one
applies high electric fields, which can lead to artefacts such as Joule heating).
Conventional laser-Doppler shift methods for assessing particle motion cannot
deconvolute diffusion and ordered motion, whereas PALS can. We have since
applied this technique to many systems in our own studies, for example to
quantify the charge on calcium carbonate particles in hydrocarbon liquids, in
work commissioned by Exxon as part their research programme into over-based
additives.[116]

Very recently, David Snoswell (from South Australia)[117] in my group has been
exploiting, in collaboration with colleagues in the electrical engineering department at Bristol and at Kodak, the use of charge-stabilised PS particles to produce
two-dimensional colloidal, photonic crystals with control of the lattice spacing. A
rotating electric field is used to generate the crystal and the spacing is controlled
by varying the field intensity. The net interaction between neighbouring particles
in the lattice is a combination of a shorter-range, repulsive electrostatic force and
a longer-range attractive dipole–dipole force, arising from the polarisation of the
electrical double layers around the particles (in other cases it might be the polarisation of the particles themselves). The applied field controls this latter
interaction, and hence the depth and position of the potential energy minimum in
which the particles reside in the crystal. Nils Elsner (from the MPI, Potsdam) and
Paddy Royal[118] have been trying to model this effect.

1.3.2   Polymers at Interfaces and Steric Interactions

Following on from the work I did at ICI, when I moved full time to Bristol in
1972, it seemed to me that studying polymer adsorption onto particles would be
a fruitful line of research to pursue. This decision was underpinned by two
important factors. The first was the movement of Tharwat Tadros, as I have
already mentioned, to ICI and his own strong interest in the steric stabilisation
of particles and droplets in agricultural formulations. It was through Tharwat
that I obtained my first research grant at Bristol: a fully funded PhD stu-
dentship by ICI (Plant Protection Division). The person appointed to the
position was Mike Garvey, who had been working in Tharwat's laboratory
and who subsequently went onto become a senior scientist at Unilever, Port
Sunlight. We decided to use various methods then available to try to determine

the thickness of adsorbed polymer layers on particles.[119-121] The polymer was poly(vinyl alcohol) (PVA), which Mike fractionated (lovingly!) into several narrow molecular weight fractions, and the techniques included sedimentation velocity (using an ultracentrifuge), electrophoresis and diffusion (using dynamic light scattering, DLS). DLS was a relatively new technique at that time (no commercial equipment was available) and we worked with two pioneering groups in the UK at that time: Peter Pusey, then at Royal Radar Establishment at Malvern, and Terry King in the physics department of Manchester University.

The second influencing factor for me to work on polymers at interfaces was the arrival in Bristol in 1973, a year after me, of a new junior fellow from Manchester University, another close friend and colleague, from that time to today: Terry Cosgrove. He was asked by Douglas Everett, following his PhD work at Manchester with Geoffrey Allen, to build a an NMR facility at Bristol, whereby physical processes such as molecular diffusion and adsorption in colloidal and interfacial systems could be studied. It seemed to Terry and I that NMR could be used to study a second feature of polymers at interfaces, namely the bound fraction of segments in polymer chains at a solid surface (*i.e.* in "trains", as opposed to "tails" or "loops").[122] We also decided to use another emerging technique at that time, small-angle neutron scattering (SANS), to investigate these systems. This should in principle lead to the segment density distribution, $\rho(z)$, normal to the interface. However, it took a very bright young PhD student, and keen mathematician, we took on from ICI at Runcorn, Trevor Crowley, to work out the mathematics involved in establishing the transforms necessary to convert the experimental $I(q)$ (I being the scattering intensity) data into $\rho(z)$ data.[123] That was a very fruitful period of research for Terry and I, and we had several joint students working on topics as diverse as the preferential adsorption of different segments of random copolymer chains on the different faces of growing wax crystals (Nigel Finch, supported by Exxon),[124] and Fourier transform infrared studies of the chemisorption of reactive silane polymers on alumina surfaces (Clive Prestidge, supported by ICI Plant Protection, and now a professor at the University of South Australia).[125,126]

It was around 1980 that Terry and I developed a much stronger collaboration with the group at Wageningen, in particular with Gerard Fleer, who had been developing, with his PhD student Jan Scheutjens, the now widely used self-consistent mean-field theory of polymers at interfaces, commonly referred to as the "Scheutjens–Fleer" theory. Another PhD student of Gerard's, Martien Cohen Stuart (now head of the Wageningen colloid group), had already been to Terry's laboratory to work on NMR. We worked with Martien, for example, on the theory of hydrodynamic flow around particles carrying adsorbed or grafted polymer layers, and showed that, at full coverage, the hydrodynamic plane of shear more or less coincides with the tail thickness.[127] In this way we were able to rationalise why the hydrodynamic thickness values that Tim Obey had previously determined using DLS for anionic poly(styrene sulfonate) chains adsorbed on cationic polystyrene latex particles were significantly greater than the r.m.s. thickness values measured by SANS;[128] although most of the chains were attached to the surface in trains through electrostatic attraction, there was still a significant fraction of segments in tails extending from the surface.

As part of the Bristol–Wageningen collaboration, joint research meetings were held every eighteen months or so; these were intended primarily as a forum where our PhD students and postdocs could present and discuss their work. The first such meeting was held in 1987 in the Ardennes in Belgium, and by the time of the fifth, in 1995, we were joined by the Lund group. Then in 1993, after some five years of "hard graft", the joint Bristol–Wageningen textbook *Polymers at Interfaces* appeared.[129]

I have continued, to the present time, to retain a strong interest in polymer adsorption. Nick Marston in my group, for example, developed a flow-through membrane device, coupled with pressure transducers, for monitoring changes in adsorbed layer thickness inside pores with changes in flow rate and solvency.[130] Also, for the last twelve years or so, a succession of students in my group (Andrew Cox, Robin Mogford, Shachi Sharma, Darby Kozak, Jess Tsipoani and Joel Manuvelpillai) have undertaken an extensive series of experiments with Lubrizol on the adsorption of fuel- and lube-oil additive polymers on model carbon particles and steel surfaces in order to understand how they function as colloidal stabilisers and anti-deposition agents.[131–135] However, the main thrust of research in this area in Bristol in the 1990s and beyond really moved to Terry's group where large strides forward have been made in using SANS and NMR techniques, in particular, to understand polymer structure and dynamics at interfaces.

One of the main reasons for trying to determine the properties (thickness, segment density distribution, *etc.*) of an adsorbed (or grafted) polymer layer on a particle surface is to better understand the role of the polymer in controlling the interparticle interactions. Provided that the polymer layer fulfils the following conditions, there will be an effective steric repulsion between the particles: (i) the adsorbed or grafted polymer layer affords sufficient coverage (*i.e.* there are no "bare patches" on the particle surface); (ii) the thickness of the polymer layer is large enough that the steric interaction is of sufficient range that the net van der Waals forces are so weak that no aggregation occurs; (iii) the stabilising chains are in a good solvent environment; and (iv) no desorption of polymer chains occurs during a particle collision.

Steric interactions offer a more "robust" repulsive interaction than electrostatic interactions, in certain environments and under certain conditions, *e.g.* non-aqueous media, high electrolyte concentrations, concentrated dispersions. In the work with Lubrizol, referred to above, in which we studied the adsorption of (relatively short-chain) polymers onto carbon particles, an open question at the beginning was: are the carbon particles electrostatically stabilised or sterically stabilised (or both)? It turns out that many of the additive polymers used have limited solubility in the various oils and are associated, both in solution and at the carbon particle surface, where they form (thick) multilayers.[134] Also the magnitude of the surface charge, measured using the PALS technique, in most cases was negligible. Hence, steric stabilisation certainly would appear to play the major role in keeping the carbon particles in dispersion.

Another, very different, example of where steric repulsion plays a more important role than electrostatic repulsion is the work carried out by Annette Murphy and Alain Hill in my group,[136] in conjunction with National Starch and Unilever, on the stability of vesicles, based on cationic, double-chain surfactants, dispersed in high concentrations of electrolyte. We synthesised a series of cationic, monodisperse polyelectrolytes, each having a terminal hydrophobic anchor unit. These hydrophobic moieties absorbed into the vesicle bilayer, leaving the polyelectrolyte chains protruding into the aqueous phase, thus providing steric stabilisation of the vesicles in the high ionic strength environment.

The examples just quoted refer to systems where strong steric stabilisation is a required condition. If any of the four conditions to achieve effective steric stabilisation, outlined earlier, is weakened, then particle aggregation may occur. Another major theme over the years in my research group has been to investigate the boundary conditions for the onset of aggregation in sterically stabilised dispersions, and the nature of the aggregation that occurs. This work started at Bristol, during my dual appointment with ICI, with my first Bristol MSc course student, John Long.[137] We decided to work on the reversible aggregation problem I had been thinking about at ICI Paints. We realised that reversibility must imply a shallow minimum in the pair interaction potential between particles. In order to mimic this situation we studied aqueous dispersions of charge-stabilised PS latex particles (of various sizes) and adsorbed a (thin) monolayer of a non-ionic surfactant ($C_{12}E_6$, where E = ethylene oxide).[138] The reason for the *thin* adsorbed layer is that the van der Waals interaction (V_A) between the particles would then be longer range than the steric interaction (V_S) afforded by the adsorbed layer. Hence, a plot of ($V_A + V_S$) *versus* the particle separation (h) would have a shallow minimum (V_{min}) in it, the depth of which would be controlled by varying the particle size. However, the particles also carried a charge, so that there was also a long-range repulsion (V_E) present between the particles. This kept the particles stable, but the reversible flocculation (*i.e.* into the shallow energy minimum) could be "switched on", by adding sufficient electrolyte to remove V_E. We also needed a method for detecting the onset of weak flocculation (using a simple spectrophotometer—not having access to angular static, let alone dynamic, light scattering at that time), and hit on the idea of using optical density (OD) measurements as a function of wavelength (λ).[137] In effect, this is an alternative method of varying the scattering vector (q), and Heller[139] had shown previously how one might relate the linear slope (n) of a plot of log(OD) *versus* log λ to the particle's size. The results of our experiments are shown in Figure 1.6, which shows plots of $n = d \log(OD)/d \log \lambda$ *versus* particle volume fraction (θ), also on a log scale, for five different sized PS particles. (Note that we use θ here for particle volume fraction, to be consistent with the original paper, but subsequently in this chapter we shall use ϕ for particle volume fraction, as we have done in more recent publications.)

Clear breaks can be seen in the plots, occurring at a critical volume fraction θ_{crit}. Below θ_{crit} the dispersions are stable, whereas above θ_{crit} reversible

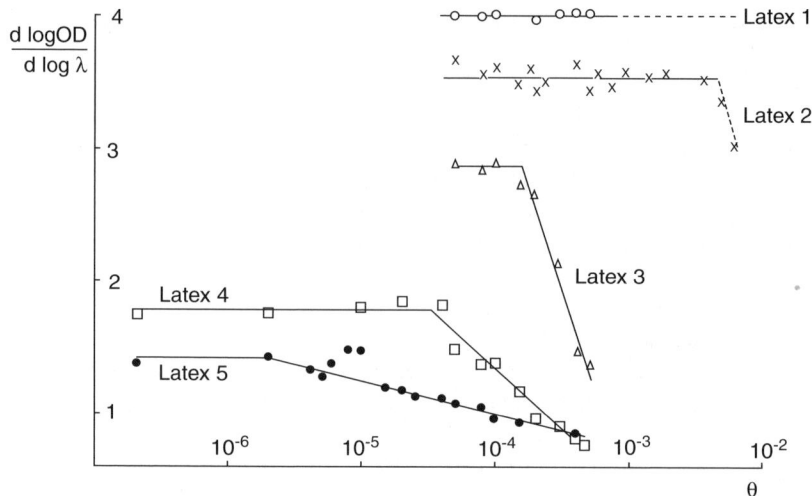

Figure 1.6   Plots of $d \log(OD)/d \log \lambda$ *versus* particle volume fraction (θ) for PS latex particles with an adsorbed layer of $C_{12}E_6$ surfactant, on adding $Ba(ClO_4)_2$ to remove any electrostatic interaction. The break points indicate the onset of weak, reversible aggregation for the different PS latex particles studied, having the following diameters: (1) 44 nm, (2) 120 nm, (3) 220 nm, (4) 500 nm, (5) 980 nm.[137]

flocculation occurs, leading to colloidal phase separation into two coexisting colloidal phases at equilibrium: a dilute (stable) phase plus a concentrated (weakly aggregated) phase. θ_{crit} decreases as the particle size increases, suggesting that the particles are flocculating into a deeper energy minimum. This process is akin to the *molecular* phase separation occurring when a vapour condenses (to a liquid or solid) above the saturation vapour pressure (at a given temperature). Molecular condensation could be described by hard-sphere perturbation theory. So, pursuing this analogy, we split[138] the free energy of aggregation (ΔG_{agg}) into two terms:

$$\Delta G_{agg} = \Delta G_{hs} + \Delta G_i \qquad (1.1)$$

where $\Delta G_{hs} (= -T\Delta S_{hs})$ is the hard sphere contribution, which is associated with the loss in translational entropy of the particles when they aggregate; since ΔS_{hs} is negative, this term opposes aggregation. ΔS_{hs} depends on $\log \phi$, and so the more dilute the dispersion the larger the ΔG_{hs} contribution. ΔG_i depends on the particle interactions. For a system where the net interaction is an attraction, as manifest in the existence of a minimum in the pair potential (V_{min}), then the deeper V_{min} the larger ΔG_i becomes. For dispersions with a fixed V_{min}, this explains the existence of a critical particle volume fraction (ϕ_{crit}) for the onset of weak aggregation and colloidal phase separation at higher ϕ values. For systems at a given value of ϕ, there will exist a critical value of V_{min} beyond which weak aggregation and colloidal phase separation occur.

Chris Cowell then embarked on a series of experiments in which we investigated the stability of both charged PS latex particles and PS-*g*-PEO particles, in the presence of PEO.[140–142] In particular, the critical flocculation temperature (CFT) of the systems was investigated, as a function of PEO concentration, particle concentration and electrolyte concentration. We were able to observe[141] that, in the two-phase coexistence region, above the CFT, the concentrated phase sometimes existed as a disordered amorphous phase, but in other cases as a pseudo-crystalline phase. The concept of constructing phase maps for weakly interacting particles began to emerge. We also made some studies of the *kinetics* of weak, reversible aggregation using a specially constructed, optically transparent Couette cell, which allowed the measurement of the optical density of the dispersion, as a function of time, at fixed shear rates.[142] In collaboration with Tharwat Tadros at ICI Plant Protection we also studied the stability of PS latex particles in the presence of adsorbed PVA (narrow molecular weight fractions)[143,144] and adsorbed PEO–poly(propylene oxide)–PEO triblock copolymers,[145] but our attention in the early 1980s largely switched to phase separation in non-aqueous dispersions.

Douglas Everett and John Stageman at Bristol had studied the stability of latex particles carrying anchored PDMS chains in hydrocarbon media[146] and, at Utrecht, Vrij's group[147] had developed a new model system: silica particles carrying terminally anchored *n*-octadecyl chains (SiO_2-*g*-C_{18} particles). Douglas and I decided to study the temperature stability of the SiO_2-*g*-C_{18} particles in various *n*-alkane ($n = 5$–7) solvents.[148] We demonstrated that these systems showed both an upper and a lower CFT. Moreover, the UCFT decreased with increasing ϕ, while the LCFT increased with increasing ϕ. The explanation for the UCFT was given in terms of the increase in V_{min} with increasing temperature, induced by the change in density and, hence, the change in Hamaker constant of the *n*-alkane solvent; the LCST was more difficult to account for. However, by constructing T *versus* ϕ phase maps we were able to predict the existence of a critical temperature in each of these systems.

An alternative explanation to account for the aggregation of particles carrying grafted chains, dispersed in a given solvent, on changing the temperature is that the (Flory) polymer–solvent χ-parameter changes, such that the mixing component of the steric interaction leads to an *attraction* between the polymer layers around the particles. This is more likely to be observed in systems where the chains are not so densely packed on the surface, such that little or no interpenetration of the chains on approaching particles can occur; one might expect this to be the case for grafted *n*-C_{18} chains. With longer, grafted *polymer* chains, however, such interpenetration is more likely to occur. In a collaboration with Jeroen van Duijneveldt, James Weeks[149] studied the effect of reducing the temperature for dispersions of SiO_2-*g*-PS particles in cyclohexane (a marginal solvent for PS). For sufficiently concentrated dispersions, as the temperature was lowered, at a given particle concentration, an instability boundary was first crossed, and then, at an even lower temperature, gel formation occurred.

Simon Emmett and Zoltan Kiraly[150] compared the stability of charge-stabilised ("Stöber") SiO_2 particles dispersed in ethanol, to which cyclohexane was added, with the stability of SiO_2-g-C_{18} particles in cyclohexane, to which ethanol was added. In the former system the aggregation was classical "DLVO" and related to changes in the zeta potential of the particles, whereas in the latter system the nature of the aggregation was similar to that reported in the previous paragraph, whereby the change in Hamaker constant of the medium is induced by the addition of ethanol, rather than a change in temperature. In collaboration with the Szeged group,[151] Keith Bean then carried out a more systematic study of the former system, determining the composite adsorption isotherm for "Stöber" SiO_2 particles in ethanol + cyclohexane mixtures, and calculated, using a Vold-type model as modified by myself,[8,9] the van der Waals attraction as a function of solvent composition. More recently Phillip Dale, in conjunction with Johan Kijlstra at Bayer in Leverkusen, has returned to study SiO_2-g-C_{18} particles, but now in water and with an adsorbed layer of various non-ionic surfactants on the surface.[152,153] Similar studies were also carried out with the monodisperse oil–water emulsions with the same surfactants.[154] Parameters controlling the CFT of the dispersions included the thickness of the surfactant layer, the particle size and the particle volume fraction. Again, theoretical calculations,[153] based on the Vincent-modified Vold model,[8,9] showed that variations in the van der Waals interactions could account for the observed CFT dependencies of the SiO_2-g-C_{18} dispersions.

Microgel particles, having surface and/or bulk charges (see Section 1.2.7) offer an interesting alternative to particles carrying grafted polymer chains as model systems for studying the interplay of longer range electrostatic and van der Waals interactions, plus short-range steric interactions (which occur on microgel particle contact).[75] In particular, with PNIPAM-based dispersions, for example, one may tune *independently* the electrostatic interaction, by varying the electrolyte concentration, and the van der Waals interaction, by changing the temperature (as the particles swell or deswell their Hamaker constant changes). Alex Routh and Mikael Rasmusson in my group[155,156] pursued this concept, in particular the fractal nature of the aggregates that formed. It was shown that, as the inter-particle attraction increased, so the fractal dimensions decreased from ~ 2.2 to ~ 1.8 (the value for diffusion-limited aggregation). The higher value of 2.2 results from the reversible nature of the aggregation, and the associated re-compaction of the aggregates, when the attraction is weak. We have also collaborated with the Almeria group on this topic.[157]

1.3.3   Depletion Interactions

There had been reports in the early literature of where polymers had been added to dispersions in order to concentrate the particles concerned by inducing colloidal phase separation, *e.g.* sodium alginate or starch to blood cells[158] or rubber latex.[159] However, the mechanism was unclear, and there was no indication of an attractive interaction being induced between the particles. I have already referred, in Section 1.1, to the early work I did at ICI on pigment particle aggregation in drying paint films, where, as the solvent evaporates, the concentration of free polymer in the continuous phase increases. To model this process we studied the flocculation kinetics of PS-g-PEO particles in water + PEO mixtures.[10] We showed that the dispersions were stable in pure water and in pure PEO, but weakly and reversibly flocculated over a certain range of PEO concentrations in water, that range being dependent on the PEO molecular weight. For potential patent reasons, publication of this work was delayed until 1975, when I was allowed to talk about it at an American Chemical Society meeting in the USA.[10]

Soon after, theoretical explanations began to emerge from various groups to try to explain the destabilising effect of adding free polymers to particulate dispersions, *e.g.* Vrij[160] in 1976, de Gennes and co-workers[161] in 1979 (using a scaling theory approach), Feigin and Napper[162] in 1980 (who coined the term "depletion" interaction) and Sperry[163] in 1981. However, all these theories were later seen to be essentially developments of a much earlier theory proposed in 1958 by Asakura and Oosawa,[164] which had largely been overlooked.

Our first publication in the "regular" literature as it were (as opposed to publication in the proceedings of a conference) in this area appeared in 1978[138] where we extended the experiments, described in the 1975 paper,[10] on PS-*g*-PEO particles in PEO + water mixtures. In 1979, de Hek and Vrij[147] published the first Utrecht experimental depletion aggregation paper; they added non-adsorbing polymers to dispersions of SiO_2-*g*-C_{18} particles in non-polar solvents. Our own first studies with non-aqueous systems were concerned with the stability of (i) cross-linked PS microgel particles[77] and (ii) SiO_2-*g*-PS particles,[141] both in PS + ethyl benzene mixtures. Previous studies on PS microgel particles with added free PS had been reported by Sieglaff.[165]

In 1984, Jan Scheutjens, Gerard Fleer and myself published[166] the first self-consistent mean-field theoretical analysis of depletion interactions, and made comparisons with some of the experimental data referred to above. By and large, good agreement was obtained. An important advance was that our theory could properly account for the "re-stabilisation" effect observed at high polymer concentrations. Feigin and Napper[162] had given a kinetic mechanism for this effect, based on the prediction of an energy barrier to aggregation occurring in the pair potential at high polymer concentrations. Our explanation,[166] in contrast, was thermodynamic: we showed that the value of V_{min} in the pair potential passes through a maximum value with increasing polymer concentration. This maximum occurs when the polymer concentration is close to the overlap concentration of the polymer chains in solution. The reason for the decrease in V_{min} above this concentration is because the depletion layer thickness itself decreases; this effect had not been considered in the earlier theories referred to above.

Other depletion flocculation studies were made by us involving particles with either grafted or adsorbed polymer layers. An early study by Jane Clarke involved SiO_2-*g*-PS particles with added free PS chains.[167] Subsequently, Jane and Ken Barnett studied PS-*g*-polyisoprene particles, prepared by an anionic dispersion polymerisation route, and polyacrylonitrile particles stabilised by adsorbed PS–PDMS–PS triblock copolymers,[168] with various added polymers. The latter system had the advantage that the coverage could be varied systematically. Later, John Edwards, Simon Emmett and Andrew Jones made further studies[169] on SiO_2-*g*-PS particles dispersed in PS + (various) solvent mixtures. In that paper[169] we developed a theory to predict the depletion interaction in dispersions of "soft spheres", *i.e.* particles carrying a grafted polymer layer. The basic idea introduced was that the free chains in solution could interpenetrate partially into the grafted chain layer, by an amount dependent on their osmotic pressure in solution and the mixing free energy of the free and grafted chains. Analytical expressions were obtained for the depletion interaction (V_{dep}), for various types of segment density distribution for the grafted layer, and also for the steric interaction (V_s). The total interaction was taken to be $V_{dep} + V_s$. The depth of the minimum in the total pair potential (V_{min}) could then be calculated under various conditions; in general, good agreement was found between the theoretical stability predictions and experimental observations. This theory was later extended by me[170] to take into account the facts that the free and grafted chains could be different, and that the grafted chains could be compressed in the presence of the free polymer. Further experimental tests of the theory, especially the role of chain coverage on the particle surface, were made by Andrew Jones and Andrew Milling.[171]

Based on a similar premise, namely that the bulk osmotic pressure of the free chains leads to compression of the free chains near the surface, I was able to derive an analytical expression for the dependence of the depletion layer thickness on bulk polymer concentration near a *hard* surface.[172] Comparisons were made with the predictions of both the de Gennes scaling theory[173] and the Scheutjens–Fleer self-consistent mean-field theory[166] for this dependence. Edwin Donath, then at the Humbolt University in Berlin, in collaboration with Greg Allan (from Melbourne University) and myself, developed and tested theoretical models to allow the calculation of the depletion layer thickness from electrophoresis[174,177] and DLS experiments.[175,176]

Predictions of experimental phase diagrams for systems with short- or long-range attractive forces (*i.e.* narrow or wide pair potential minima, induced by either van der Waals or depletion attraction forces), and the existence of "solid–vapour", "liquid–vapour" and "liquid–solid" molecular analogues, were made by our group.[178,179] To this end relatively simple statistical thermodynamic mean-field models of each phase were used, and phase equilibria predicted by equating the chemical potentials and also the pressures in coexisting phases.[179] Figure 1.7 illustrates coexisting (fluid–solid and gas–liquid) colloidal phases for SiO_2-g-C_{18} particles in mixtures of PDMS + cyclohexane.[179]

Further experimental work from our group on depletion flocculation followed. For example, Paul Jenkins[180] studied the depletion flocculation of SiO_2-g-C_{18} particles in binary mixtures of polymers, where the mixtures were either PS or PDMS of two molecular weights, or mixtures of PS and PDMS. Explanations of the observed effects required the occurrence of *non-equilibrium* behaviour.

Following on from James Weeks' work,[149] described in the previous section, Gerrit Vliegenthart, Jeroen van Duijneveldt and myself studied dispersions of SiO_2-g-PS particles in mixtures of PS and benzene.[181] We were able to change independently the short-range (steric) and longer range (depletion) interactions, by changing the temperature or the free PS concentration, respectively. We showed the combination of the two-phase transition.

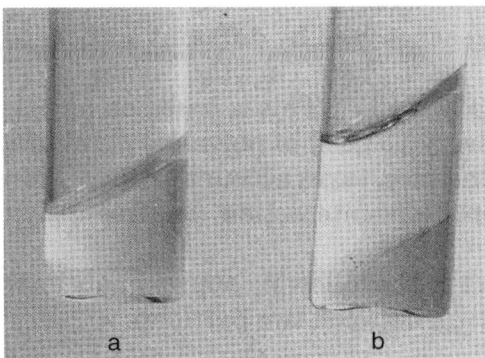

Figure 1.7 Example of coexisting colloidal phases for the system SiO_2-g-C_{18} + cyclohexane, at 20 °C: (a) fluid–solid; (b) gas–liquid. The tubes have been tilted to show the fluidity of the interface.[179]

Andrew Milling and Nick Cawdery[182] studied, respectively, dispersions of (i) SiO_2-g-PMMA particles in PMMA + dioxan mixtures and (ii) PS-g-PEO particles in PAAc + water mixtures. In the former system the effects of the amount of PMMA grafted and molecular weight on the depletion interaction were systematically investigated. In the latter system, at low pH, where the PAAc chains are in their neutral, undissociated state, they "adsorb" onto the PEO chains grafted to the PS particles, and classical *bridging* flocculation was observed at low PAAc concentrations. At high pH, however, where the PAAc chains are in the anionic, polyelectrolyte form, then no adsorption is observed, and *depletion* flocculation was observed at higher PAAc concentrations. In a later paper, Milling compared the destabilisation of silica particles in the presence of PAAc in water to the force–separation curves for this system, measured using an atomic force microscope.[183]

More recently, Ruth Dunleavey[184] also studied the depletion aggregation of silica particles, in this case in the presence of sodium poly(styrene sulfonate) and KCl. She found that, under certain conditions, aggregation rate constants much faster than the corresponding (Smoluchoski) diffusion-controlled rate constant were observed. The explanation given was in terms of the long-range nature of the depletion interaction; this suggestion is backed up by the fact that the rate constants approached the diffusion-controlled limit when the dispersions were diluted sufficiently.[184]

1.3.4 Interactions in Mixed Particle Systems

Another early research interest I developed at Bristol concerned the interaction of small particles of one charge sign with much larger particles of the opposite sign, and to see how layers of adsorbed polymer on each set of particles would moderate the interaction. In effect we were mimicking, with colloidal particles, molecular *adsorption*, induced by electrostatic attraction in this case, in a somewhat similar way that colloidal phase separations, as described in the previous sections, mimicked molecular *vapour condensation*, induced by van der Waals forces or depletion attraction. The main reasons for using larger particles as the substrate, rather than flat surfaces (see Section 1.3.5), were that (i) the specific surface area of the adsorbent is much larger and (ii) particle adsorption equilibria (and kinetics) could be readily followed by separating, by centrifugation, the large particles from the small particles remaining in the non-adsorbed state, and measuring the concentration of the latter, say by turbidiometry.

Another of my early PhD students at Bristol, Colin Young (later, after marriage, Mumme-Young), carried out the first experiments in this area.[185,186] A typical set of particle adsorption isotherms is shown in Figure 1.8, for different background NaCl concentrations. The adsorbed amount is expressed as the coverage, defined as $\theta = \Gamma/\Gamma_{hcp}$, where Γ is the adsorbed amount and Γ_{hcp} is the maximum adsorbed amount, corresponding to two-dimensional, hexagonal close-packing of the adsorbed particles on the surface. Here the large negative PS particles, used as the substrate, were 3.2 μm in diameter, and the small positive, adsorbing PS particles were 0.2 μm. Both sets of particles carried an adsorbed layer of PVA (molecular weight 10 500) to provide a steric repulsive interaction. The particle adsorption isotherms change shape with increasing NaCl concentration. There are three basic forms. At low NaCl concentrations (10^{-5} and 10^{-3} M) the isotherms are high affinity, but the plateau level reached is higher at the higher salt concentration. The reason for this is that there are two interactions to consider: (i) the normal interaction between the large and

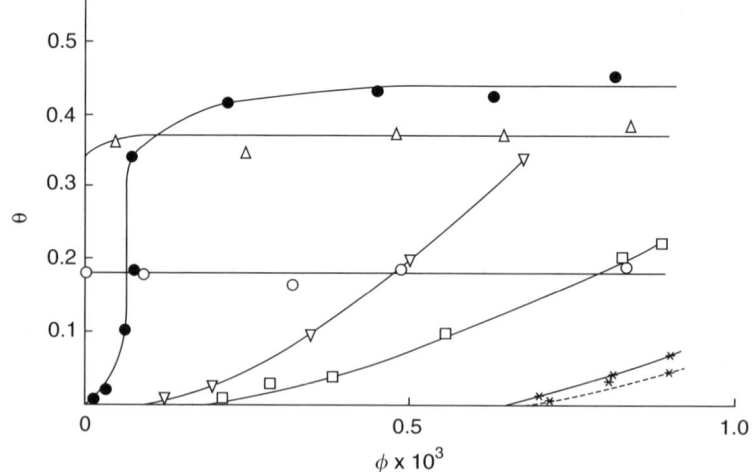

Figure 1.8   Particle adsorption isotherms for small (0.2 μm) cationic PS particles onto larger (3.2 μm) anionic PS particles, where each set of particles carries a thin adsorbed layer of PVA (molecular weight 10 500), as a function of NaCl concentration: $\bigcirc$, 10^{-5} M; $\triangle$, 10^{-3} M; $\bullet$, 3×10^{-3} M; $\triangledown$, 10^{-2} M; $\square$, 10^{-1} M; *, 1 M. The dashed line indicates the adsorption of small *anionic* PS particles at 1 M NaCl.[185]

small particles (this is always attractive and leads to strong adsorption) and (ii) the lateral interaction between the adsorbed small particles on the surface (this is repulsive, but becomes weaker at higher salt concentrations). This latter effect allows closer packing on the surface. The second isotherm form is that occurring at high NaCl concentrations (10^{-2} M and greater). These are low affinity ("type III" in gas adsorption terms). We showed that in this region two-dimensional clusters or "rafts" of small particle aggregates formed on the surface, because the lateral interaction between neighbouring adsorbed particles, at these higher NaCl concentrations, has become an attraction, rather than a repulsion. At the highest NaCl concentration studied (1 M) the adsorption is very weak, and the adsorption of the positive particles is essentially the same as for small *negative* particle (of similar size) onto the larger particles. This suggests that, under these conditions, both the normal and lateral attractions are just van der Waals attractions, offset by the steric interaction, resulting in very shallow energy minima in both interactions. The third isotherm shape in Figure 1.8 occurs at 3×10^{-3} M NaCl, *i.e.* at an intermediate NaCl concentration. This can be described as a transition region between the high- and low-affinity adsorption regions; it occurs at a specific (or over a narrow range of) NaCl concentrations, c^*. In some respects, c^* resembles the classical critical electrolyte concentration for coagulation in charge-stabilised dispersions; c^* decreases with increasing PVA molecular weight. Bill van Megen and Ian Snook, at RMIT in Melbourne, undertook some successful Monte Carlo modelling of our systems.[187] Later Steve Harley in my group, working with Dudley Thompson, then our electron microscopist in Bristol, produced some excellent scanning electron microscopy (SEM) images of small particles adsorbed on big particles, some of which are shown in Figure 1.9, and successfully modelled the kinetics of particle adsorption.[188]

Paul Luckham, who now holds a chair at Imperial College, then took up this theme as his PhD project with me.[189–193] As well as exploring the kinetics of particle adsorption in very similar systems to those studied earlier,[189] we also studied the effect of temperature on the adsorption isotherms, and showed that, unlike most cases of molecular adsorption, in our systems the adsorbed amount of particles (with adsorbed PVA layers) increased with increasing temperature, because the depth of the minimum in the pair potential increased with temperature, as the adsorbed PVA layers became thinner.[189] Paul then went on to study particle bridging flocculation in these systems. Various techniques were employed including sedimentation,[190] freeze-fracture SEM[191] and rheology.[192] This phenomenon is akin to polymer bridging flocculation, and is optimal at around half-coverage of the large particles by the small ones, as indicated, for example, by a maximum in the yield stress of the system at this point.[192] Nick Marston and Nigel Wright later made similar studies of the adsorption of monodisperse, anionic titanium dioxide particles on cationic latex particles.[194]

Tharwat Tadros had been involved as the industrial sponsor in the work of Colin Young and Paul Luckhams, and he challenged me to see if one could also observe similar particle adsorption and bridging phenomena in *non-aqueous* systems. David Skuse worked successfully on this problem.[195] Instead of

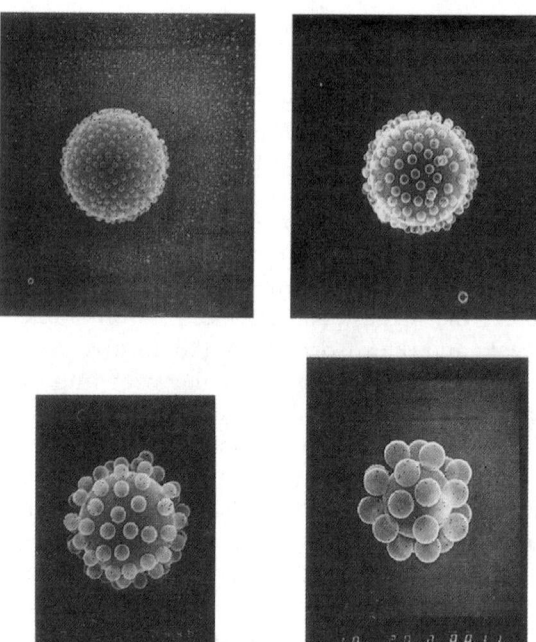

Figure 1.9   Thin-film, freeze-drying SEM images of small, cationic PS particles of increasing size, adsorbing onto a large, anionic PS particle.[188]

opposite charge interactions, we decided to use Lewis acid–base, dipolar interactions between two sets of silica particles dispersed in toluene. This was achieved by adsorbing onto one set of silica particles a copolymer containing weak acid (methacrylic acid, MAAc) groups, and onto the other set a co-polymer containing weak base (dimethylaminoethyl methacrylate, DMEAM) groups. On mixing the two sets of particles, weak flocculation was induced over a certain range of particle concentration ratios. Moreover, if a *strong* base (*e.g.* *n*-propylamine) or a *strong* acid (*e.g. p*-toluene sulfonic acid) was then added, the system deflocculated, as these displaced the weak acid–base interactions between the two adsorbed copolymers.[195]

We have also investigated mixed emulsion systems. Robin Davies, in a project funded by BP with David Graham,[196,197] looked at the interaction between two sets of water droplets, mixed in an oil continuous phase, where one set of droplets contained pure water and the other high concentrations of NaCl. Above a critical NaCl concentration, hetero-coalescence was induced, but no homo-coalescence. This was thought to be due to the strong osmotic gradient set up across the thin oil film between two contacting unlike droplets. More recently Phillip Dale[154] extended the CFT studies he had carried out on surfactant-stabilised, monodisperse oil–water emulsions, referred to at the end of Section 1.3.2, to investigate effects occurring in *binary mixtures* of oil droplets, each with a different surfactant layer, in water; both Ostwald ripening and surfactant exchange were observed.

Mixed polymer latex particles have been widely used to study different aspects of hetero-flocculation. We have also explored these systems. For example, Pierre Starck,[198] essentially extending the earlier work of Nick Cawdery[182] referred to in Section 1.3.3, investigated, using rheological techniques (primarily a strain relaxation method to determine yield stresses), the hetero-aggregation between two sets of latex core–shell particles, one having a shell of MAAc groups and the other a shell of PEO groups. The primary interparticle interaction here is hydrogen bonding (at low pH) between the hydrogen atoms on the –COOH groups and the oxygen atoms on the PEO chains.

Alex Routh studied the hetero-flocculation kinetics between two sets of oppositely charged microgel particles, PNIPAM and PVP, and showed that the presence of Na^+ ions led to some unexpected anomalous behaviour.[199] Previously, together with the Almeria group, we had looked at hetero-flocculation in mixtures of hard anionic PS particles and soft (swollen), cationic PVP microgel particles, using both static and dynamic light scattering to investigate flocculation kinetics and floc morphology.[200,201] Van der Waals interactions do not contribute to the hetero-flocculation, only electrostatic interactions. The flocs were much more "open-network" than predicted from pure diffusion-limited aggregation, because of the presence of electrostatic repulsions between the like-charge particles present within the floc. David Snoswell, in a project funded by Kodak,[202] later exploited this effect to produce highly porous aggregates of controlled, variable porosity (and hydrophobicity) for potential use as an absorbent layer in ink-jet printing papers. David has also recently investigated the formation of long, flexible particle "chains" by hetero-aggregating, in electric fields, mixtures of anionic PS particles and cationic PVP particles.[203]

1.3.5   Deposition of Particles on Flat Surfaces

As mentioned at the beginning of the previous section, the deposition of small particles on much larger particles offers a convenient way of studying particle adsorption phenomena. Similar adsorption behaviour on *flat* surfaces normally requires observation of the adsorbed layer itself, by some microscopic or scattering/reflectivity technique. Our first work in this area was that of Nick Marston,[204] who used an optical microscope, in combination with a stagnant-point flow cell, to study the kinetics of deposition of titanium dioxide particles onto cationically modified silica surfaces. He was also able to study the phenomenon of two-dimensional particle "raft formation" we had observed in the earlier work of Colin Young,[185,186] referred to in the previous section.

Kathryn Rees, in work funded by Unilever, again using stagnant-point flow and other types of flow cells, recently[205] investigated the deposition of our monodisperse PDMS-based droplets and microgel particles (see Section 1.2.5 and refs 58–60) on optically flat substrates having controlled surface properties (charge and degree of hydrophobicity). In particular, she has looked at the adhesion and spreading of the oil droplets on the various surfaces.

Verawan Nerapusri in my group has studied, in collaboration with Joe Keddie at Surrey University,[206] the swelling properties of polyNIPAM and poly(NIPAM-*co*-AAc) microgel particle monolayers, deposited on oxidised silicon surfaces, using spectral ellipsometry. In general, the layer thickness response to temperature changes directly mirrored that of the diameter of the free microgel particles in dispersion, but changes in response to pH changes were more subtle, because of strong electrostatic interactions between the microgel particles and the substrate (a cationic silica surface). We have also looked at the interaction of cationic surfactants with similar microgel monolayers.[111]

Paul Davies, in a collaboration with Adam Feiler at Uppsala University, has recently been investigating the uptake of gold nanoparticles into PVP microgel particles, deposited on surfaces, using a quartz crystal microbalance and an atomic force microscope.[113] He has been able to follow the swelling/ deswelling of the PVP layer with changes in pH, as well as the uptake of the gold particles.

1.4   Concluding Remarks

Very few academics can be successful these days on their own. They owe much to their co-workers, their collaborators in other laboratories and of course their sponsors. This is very true in my case, and in this article I have tried to acknowledge most of the individuals and company-based sponsors without whom I would have achieved very little. If I have inadvertently missed anyone, then please accept my apologies!

There are other organisations which have helped "colour" my career, which I would like to acknowledge also. These are the learned, professional societies. From my early days on the staff at Bristol I served on the committee of the colloid and surface chemistry group of the Society of Chemical Industry (SCI), later becoming chair of that group. Subsequently, I served on the committee of the colloid and interface science group of the Royal Society of Chemistry (RSC), becoming chair in due course of that group as well. One consequence of having served as chair of both of these groups was the move I initiated to have the two committees amalgamate to form a joint colloid and surface group in the UK. I have also enjoyed working on the committees of the Rideal Trust and of the UK Polymer Colloids Forum (UKPCF). I am also pleased to have been involved with helping to organise many colloid PhD student meetings over the years, starting in 1987 with the Wageningen group, as I mentioned earlier, and subsequently on a wider European basis with the SCI/RSC groups, and now with the European Colloid and Interface Science Group (ECIS). On the international front, apart from all the research collaborations with many friends throughout the world, in particular in Europe and Australia, I am perhaps proudest of all to have served (2003–06) as president of the International Association of Colloid and Interface Scientists. So to all my many colloid friends and colleagues, in Bristol and around the world, many, many thanks!

References

1. A.J. Leadbetter, D.J. Taylor and B. Vincent, *Can. J. Chem.*, 1964, **42**, 2930–2932.
2. B. Vincent, in *Proc. 6th Int. Congress on Surface Active Agents*, Hanser-Verlag, 1973, part B,pp. 581–592.
3. R.H. Ottewill and B. Vincent, *J. Chem. Soc. Faraday Trans. 1*, 1972, **68**, 1533–1543.
4. B. Vincent, *Adv. Colloid Interface Sci.*, 1992, **42**, 279–302.
5. B. Vincent and J. Lyklema, *Spec. Discuss. Faraday Soc.*, 1971, **1**, 148–157.
6. B. Vincent, B.H. Bijsterbosch and J. Lyklema, *J. Colloid Interface Sci.*, 1971, **37**, 171–178.
7. M.J. Vold, *J. Colloid Interface Sci.*, 1961, **16**, 1.
8. D.W.J. Osmond, B. Vincent and F.A. Waite, *J. Colloid Interface Sci.*, 1973, **42**, 262–269.
9. B. Vincent, *J. Colloid Interface Sci.*, 1973, **42**, 270–285.
10. F.K.R. Li-In-On, B. Vincent and F.A. Waite, *Am. Chem. Soc. Symp. Ser.*, 1975, **9**, 165–172.
11. B.V. Derjaguin and L. Landau, *Acta Phys. Chim URSS*, 1941, **14**, 633.
12. E.J.W. Verwey and J.Th.G. Overbeek, *Theory of the Stability of Lyophobic Colloids*, Elsevier, 1948.
13. J.M. Saunders, J.W. Goodwin, R.M. Richardson and B. Vincent, *J. Phys Chem.*, 1999, **103**, 9211–9218.
14. T.H. Muster and B. Vincent, *Colloids Surf. A*, 2003, **228**, 181–187.
15. D. Voisin and B. Vincent, *Adv. Colloid Interface Sci.*, 2003, **106**, 1–22.
16. C. Martin, PhD thesis, University of Bristol, 2007.
17. A. Olsen, H.C. Lee, M. Hatzopoulos, J.S. van Duijneveldt, and B. Vincent, in preparation.
18. P.S. Bolt, J.W. Goodwin and R.H. Ottewill, *Langmuir*, 2005, **21**, 9911–9916.
19. A.R. Hemsley, M.E. Collinson, W.I. Kovach, B. Vincent and T. Williams, *Philos. Trans. R. Soc. London, Ser. B*, 1994, **345**, 163–173.
20. A. Hemsley, P. Jenkins, M. Collinson and B. Vincent, *Botanical J. Linnean Soc.*, 1996, **121**, 177–187.
21. A.R. Hemsley, B. Vincent, M.E. Collinson and P.C. Griffiths, *Ann. Botany*, 1998, **82**, 105–109.
22. A.R. Hemsley, M.E. Gollinson, B. Vincent, P.C. Griffiths and P.D. Jenkins, in *Pollen and Spores: Morphology and Biology*, ed. M.R. Harley, C.M. Morton and S. Blackmore, Royal Botanical Gardens, Kew, 2000, pp. 31–44.
23. D. Fairhurst, K. Bridger and B. Vincent, *J. Colloid Interface Sci.*, 1979, **68**, 190–195.
24. J. Clarke and B. Vincent, *J. Colloid Interface Sci.*, 1981, **82**, 208–216.
25. K. Bridger and B. Vincent, *Eur. Polym. J.*, 1980, **16**, 1017–1021.
26. J. Edwards, S. Lennon, A.F. Toussaint and B. Vincent, *Am. Chem. Soc. Symp. Ser.*, 1984, **240**, 281–298.
27. C. Bromley, *Colloids Surf.*, 1989, **17**, 1.

28. C. Cowell and B. Vincent, *J. Colloid Interface Sci.*, 1982, **87**, 518–526.
29. K. Ryan, *Chem. Ind.*, 1988, 359–364.
30. P. Pendleton and B. Vincent, in *Proc. 5th London International Carbon and Graphite Conference*, Society of Chemical Industry, London, 1978, p. 452.
31. P. Pendleton, B. Vincent and M.L. Hair, *J. Colloid Interface Sci.*, 1981, **80**, 512–527.
32. J. Edwards, R. Fisher and B. Vincent, *Makromol. Chem.*, 1983, **4**, 393–397.
33. S.P. Armes, B. Vincent and J.W. White, *J. Chem. Soc., Chem. Commun.*, 1986, 1525–1527.
34. S.P. Armes and B. Vincent, *J. Chem. Soc., Chem. Commun.*, 1987, 288–290.
35. S.P. Armes, J.F. Miller and B. Vincent, *J. Colloid Interface Sci.*, 1987, **118**, 410–416.
36. N. Cawdery, T.M. Obey and B. Vincent, *J. Chem. Soc., Chem. Commun.*, 1988, 1189–1190.
37. G. Markham, T.M. Obey and B. Vincent, *Colloids Surf.*, 1990, **51**, 239–254.
38. J.Th.G. Overbeek, in *Colloid Science*, ed. H.R. Kruyt, Elsevier, 1952, p. 209.
39. E.C. Cooper and B. Vincent, *J. Colloid Interface Sci.*, 1989, **132**, 592–594.
40. E.C. Cooper and B. Vincent, *J. Phys. D: App. Phys.*, 1989, **22**, 1580–1585.
41. B. Vincent and J. Waterson, *J. Chem. Soc., Chem. Commun.*, 1990, 683–684.
42. J.W. Goodwin, G. Markham and B. Vincent, *J. Phys. Chem. B*, 1997, **101**, 1961–1967.
43. M.C. Barker and B. Vincent, *Colloids Surf.*, 1984, **8**, 289–296.
44. M.C. Barker and B. Vincent, *Colloids Surf.*, 1984, **8**, 297–314.
45. Z. Kiraly and B. Vincent, *Polym. Int.*, 1992, **28**, 139–150.
46. S. Biggs and B. Vincent, *Colloid Polym. Sci.*, 1992, **270**, 505–510.
47. S. Biggs and B. Vincent, *Colloid Polym. Sci.*, 1992, **270**, 511–517.
48. S. Biggs and B. Vincent, *Colloid Polym. Sci.*, 1992, **270**, 563–573.
49. A. Horgan, B. Saunders, B. Vincent and R.K. Heenan, *J. Colloid Interface Sci.*, 2003, **262**, 548–559.
50. A. Horgan and B. Vincent, *J. Colloid Interface Sci.*, 2003, **262**, 536–547.
51. B. Binks, (ed.), *Modern Aspects of Emulsion Science*, Royal Society Chemistry, 1998.
52. R. Hengelmolen and B. Vincent, *J. Chem. Soc., Faraday Trans.*, 1997, **93**, 3683–3688.
53. R. Hengelmolen, B. Vincent and G. Hassall, *J. Colloid Interface Sci.*, 1997, **196**, 12–22.
54. P.J. Dowding, J.W. Goodwin and B. Vincent, *Colloids Surf.*, 1998, **145**, 263–270.
55. P.J. Dowding and B. Vincent, *Colloids Surf.*, 2000, **161**, 259–269.
56. P.J. Dowding, J.W. Goodwin and B. Vincent, *Colloid Polym. Sci.*, 2000, **278**, 346–351.
57. P.J. Dowding, J.W. Goodwin and B. Vincent, *Colloids Surf.*, 2001, **180**, 301–309.
58. T.M. Obey and B. Vincent, *J. Colloid Interface Sci.*, 1994, **163**, 454–463.
59. K. Anderson, T.M. Obey and B. Vincent, *Langmuir*, 1994, **10**, 2493–2494.

60. B. Neumann, B. Vincent, R. Krustev and H.J. Müller, *Langmuir*, 2004, **20**, 4336–4344.
61. W. Stöber, A. Fink and E. Bohn, *J. Colloid Interface Sci.*, 1968, **26**, 62.
62. M.I. Goller, T.M. Obey, D.O.H. Teare, B. Vincent and M. Wegener, *Colloids Surf.*, 1997, **123/124**, 183–193.
63. B. Vincent, Z. Kiraly and T.M. Obey, in *Modern Aspects of Emulsion Science*, ed. B. Binks, Royal Society Chemistry, 1998, pp. 100–114.
64. Z. Kiraly and B. Vincent, *Polym. Int.*, 1992, **28**, 139–150.
65. M.I. Goller and B. Vincent, *Colloids Surf.*, 1998, **142**, 281–285.
66. M. O'sullivan, PhD thesis, University of Bristol, 2007.
67. A. Loxley and B. Vincent, *J. Colloid Interface Sci.*, 1998, **208**, 49–62.
68. P.J. Dowding, R. Atkin, B. Vincent and P. Bouillot, *Langmuir*, 2004, **20**, 11374–11379.
69. P.J. Dowding, R. Atkin, B. Vincent and P. Bouillot, *Langmuir*, 2005, **21**, 5278–5284.
70. M.S. Romero-Cano and B. Vincent, *J. Controlled Release*, 2002, **82**, 127–135.
71. H. Wassenius, M. Nydén and B. Vincent, *J. Colloid Interface Sci.*, 2003, **264**, 538–547.
72. R. Atkin, P. Davies, J. Hardy and B. Vincent, *Macromolecules*, 2004, **37**, 7979–7985.
73. B.R. Saunders and B. Vincent, *Adv. Colloid Interface Sci.*, 1999, **80**, 1–25.
74. B.R. Saunders and B. Vincent, in *Encyclopedia of Surface and Colloid Science*, Marcel Dekker, New York, 2002, pp. 4544–4559.
75. B. Vincent and B. Saunders, in *Colloid Stability: II. The Role of Surface Forces*, ed. Th.F. Tadros, Wiley-VCH, 2007, pp. 183–202.
76. R.H. Pelton and P. Chibante, *Colloids Surf.*, 1986, **120**, 247.
77. J. Clarke and B. Vincent, *J. Chem. Soc., Faraday Trans. 1*, 1981, **77**, 1831–1843.
78. M.J Snowden, B. Vincent and J.C. Morgan, *UK Pat.*, GB2262117A, 1993.
79. M.J. Snowden and B. Vincent, *J. Chem. Soc., Chem. Commun.*, 1992, 1103–1105.
80. M.J. Snowden and B. Vincent, *ACS Symp. Ser.*, 1993, **532**, 153–160.
81. M.J. Snowden, N. Marston and B. Vincent, *Colloid Polym.*, 1994, **272**, 1273–1280.
82. P.J. Dowding, B. Vincent and E. Williams, *J. Colloid Interface Sci.*, 2000, **221**, 268–272.
83. B. Sierra-Martin, M.S. Romero-Cano, T. Cosgrove, B. Vincent and A. Fernandez-Barbero, *Macromolecules*, 2005, **38**, 10782–10787.
84. B.R. Saunders, H.M. Crowther, G.E. Morris, S.J. Mears, T. Cosgrove and B. Vincent, *Colloids Surf.*, 1999, **149**, 57–64.
85. H.M. Crowther, B.R. Saunders, S.J. Mears, T. Cosgrove, B. Vincent, S.M. King and G.E. Yu, *Colloids Surf.*, 1999, **152**, 327–331.
86. H.M. Crowther and B. Vincent, *Colloid Polym. Sci.*, 1998, **276**, 46–51.

87. B.R. Saunders, H. Crowther and B. Vincent, *Macromolecules*, 1997, **30**, 482–487.
88. P. Bouillot and B. Vincent, *Colloid Polym. Sci.*, 2000, **278**, 74–79.
89. M. Rasmusson, B. Vincent and N. Marston, *Colloid Polym. Sci.*, 2000, **278**, 253–258.
90. A. Loxley and B. Vincent, *Colloid Polym. Sci.*, 1997, **275**, 1108–1114.
91. M.J. Snowden, B. Chowdery, B. Vincent and G. Morris, *J. Chem. Soc., Faraday Trans.*, 1996, **92**, 5013–5016.
92. S. Neyret and B. Vincent, *Polymer*, 1997, **38**, 6129–6134.
93. M. Bradley, B. Vincent and G. Burnett, *Aust. J. Chem.*, in press.
94. A. Fernández-Nieves, A. Fernández-Barbero, B. Vincent and F.J. de las Nieves, *Macromolecules*, 2000, **33**, 2114–2118.
95. A. Fernández-Nieves, A. Fernández-Barbero, B. Vincent and F.J. de las Nieves, *J. Chem. Phys.*, 2003, **119**, 10385–10388.
96. A. Fernández-Nieves, A. Fernández-Barbero, F.J. de las Nieves and B. Vincent, *J. Phys.: Condens. Matter*, 2000, **12**, 3605–3614.
97. M. Bradley, J. Ramos and B. Vincent, *Langmuir*, 2005, **21**, 209–215.
98. R. Atkin, M. Bradley and B. Vincent, *Soft Matter*, 2005, **1**, 160–165.
99. I. Kaneda and B. Vincent, *J. Colloid Interface Sci.*, 2004, **274**, 49–54.
100. M. Bradley, B. Vincent and G. Burnett, *Langmuir*, in press.
101. M.J. Snowden, D. Thomas and B. Vincent, *Analyst*, 1993, **118**, 1367–1369.
102. G. Morris, B. Vincent and M.J. Snowden, *J. Colloid Interface Sci.*, 1997, **190**, 198–205.
103. G. Morris, B. Vincent and M.J. Snowden, *Prog. Colloid Polym. Sci.*, 1997, **105**, 16–22.
104. B.R. Saunders and B. Vincent, *Colloid Polym. Sci.*, 1997, **275**, 9–17.
105. B.R. Saunders and B. Vincent, *J. Chem. Soc., Faraday Trans.*, 1996, **92**, 3385–3389.
106. B.R. Saunders and B. Vincent, *Prog. Colloid Polym. Sci.*, 1997, **105**, 11–15.
107. A. Routh, A. Fernández-Nieves, M. Bradley and B. Vincent, *J. Phys. Chem. B*, 2006, **110**, 12721–12727.
108. M. Bradley and B. Vincent, *Langmuir*, 2005, **21**, 8630–8634.
109. V. Nerapusri, J.L. Keddie, B. Vincent and I. Bushnak, *Langmuir.*, 2006, **22**, 5036–5041.
110. M. Bradley and B. Vincent, *Langmuir*, in press.
111. V. Nerapusri, J.L. Keddie, B. Vincent and I.A. Bushnak, *Langmuir*, in press.
112. M. Bradley, N. Bruno and B. Vincent, *Langmuir*, 2005, **21**, 2750–2753.
113. P. Davies, PhD thesis, University of Bristol, 2008; P. Davies, A. Feiler and B. Vincent, in preparation.
114. D. Wyatt and B. Vincent, *J. Biopharmaceut. Sci.*, 1993, **3**, 27–31.
115. J.F. Miller, K. Schätzel and B. Vincent, *J. Colloid Interface Sci.*, 1991, **143**, 532–554.
116. J.F. Miller, B.J. Clifton, P.R. Benneyworth, B. Vincent, I.P. MacDonald and J.F. Marsh, *Colloids Surf.*, 1992, **66**, 197–202.
117. D.R.E. Snoswell, C.L. Bower, P. Ivanov, M.J. Cryan, J.G. Rarity and B. Vincent, *New J. Phys.*, 2006, **8**, 267–276.

118. N. Elsner, P. Royal, D.E. Snoswell and B. Vincent, in preparation.
119. M.J. Garvey, Th.F. Tadros and B. Vincent, *J Colloid Interface Sci*, 1974, **49**, 57–68.
120. M.J. Garvey, Th.F. Tadros and B. Vincent, *J. Colloid Interface Sci.*, 1976, **55**, 440–453.
121. Th. van den Boomgaard, T.A. King, Th.F. Tadros, H. Tang and B. Vincent, *J. Colloid Interface Sci.*, 1978, **66**, 68–76.
122. K.G. Barnett, T. Cosgrove, B. Vincent, M. Cohen-Stuart and D.S. Sissons, *Macromolecules*, 1981, **14**, 1018–1020.
123. T. Cosgrove, T.L. Crowley, B. Vincent, K.G. Barnett and Th.F. Tadros, *J. Chem. Soc., Faraday Symp.*, 1981, **161**, 101–108.
124. T. Cosgrove, N. Finch, B. Vincent and J. Webster, *Colloids Surf.*, 1988, **31**, 33–46.
125. T. Cosgrove, C.A. Prestidge and B. Vincent, *J. Chem. Soc., Faraday Trans.*, 1990, **86**, 1377–1382.
126. T. Cosgrove, C.A. Prestidge, S.M. King and B. Vincent, *Langmuir*, 1992, **8**, 2206–2209.
127. M.A. Cohen-Stuart, F.H.L.H. Waajen, T. Cosgrove, B. Vincent and T. Crowley, *Macromolecules*, 1984, **17**, 1825–1830.
128. T. Cosgrove, T.M. Obey and B. Vincent, *J. Colloid Interface Sci.*, 1986, **111**, 409–418.
129. G.J. Fleer, M.A. Cohen Stuart, J.M.H.M. Scheutjens, T. Cosgrove and B. Vincent, *Polymers at Interfaces*, Chapman and Hall, 1993.
130. N.J. Marston and B. Vincent, *Colloids Surf.*, 1998, **141**, 73–79.
131. A.R. Cox, B. Vincent, S. Harley and S.E. Taylor, *Colloids Surf. A*, 1999, **146**, 153–162.
132. A.R. Cox, R. Mogford, B. Vincent and S. Harley, *Colloid Surf. A*, 2001, **181**, 205–213.
133. S.D. Gurumayum Sharma, D. Moreton and B. Vincent, *Colloids Surf. A*, 2002, **210**, 139–149.
134. S.D. Gurumayum Sharma, D. Moreton and B. Vincent, *J. Colloid Interface Sci.*, 2003, **263**, 343–349.
135. S.D. Gurumayum Sharma, H. Alford, D. Moreton and B. Vincent, *Colloids Surf. A*, 2004, **250**, 51–56.
136. A. Murphy, A. Hill and B. Vincent, *Berg. Bunsenges Ges. Phys. Chem.*, 1996, **100**, 963–971.
137. J.A. Long, D.W.J. Osmond and B. Vincent, *J. Colloid Interface Sci.*, 1973, **42**, 545–553.
138. C. Cowell, R. Li-In-On and B. Vincent, *J. Chem. Soc., Faraday Trans. 1*, 1978, **74**, 337–347.
139. W. Heller and W.J. Pangonis, *J. Chem. Phys.*, 1957, **26**, 498.
140. C. Cowell and B. Vincent, in *The Effect of Polymers on Dispersion Properties*, ed. Th.F. Tadros, Academic Press, 1982, pp. 263–284.
141. C. Cowell and B. Vincent, *J. Colloid Interface Sci.*, 1982, **87**, 518–526.
142. C. Cowell and B. Vincent, *J. Colloid Interface Sci.*, 1983, **95**, 573–582.

143. R. Lambe, Th.F. Tadros and B. Vincent, *J. Colloid Interface Sci.*, 1978, **66**, 77–84.

144. Th.F. Tadros and B. Vincent, *J. Colloid Interface Sci.*, 1979, **72**, 505–514.

145. Th.F. Tadros and B. Vincent, *J. Phys. Chem.*, 1980, **84**, 1575–1580.

146. D.H. Everett and J.F. Stageman, *Colloid Polym. Sci.*, 1977, **255**, 293.

147. H. de Hek and A. Vrij, *J. Colloid Interface Sci.*, 1979, **70**, 552.

148. J. Edwards, D.H. Everett, T. O'sullivan, I. Pangalou and B. Vincent, *J. Chem. Soc., Faraday Trans. 1*, 1984, **80**, 2599–2607.

149. J.R. Weeks, J.S. van Duijneveldt and B. Vincent, *J. Phys.: Condens. Matter*, 2000, **12**, 9599–9606.

150. B. Vincent, Z. Kiraly, S. Emmett and A. Beaver, *Colloids Surf.*, 1990, **49**, 121–132.

151. Z. Kiraly, L. Turi, I. Dekany, K. Bean and B. Vincent, *Colloid Polym. Sci.*, 1996, **274**, 779–787.

152. P. Dale, B. Vincent, T. Cosgrove and J. Kijlstra, *Langmuir*, 2005, **21**, 12244–12249.

153. P. Dale, J. Kilstra and B. Vincent, *Langmuir*, 2005, **21**, 12250–12256.

154. P. Dale, J. Kijlstra and B. Vincent, *Colloids Surf. A*, 2006, **291**, 85–92.

155. A. Routh and B. Vincent, *Langmuir*, 2002, **18**, 5366–5369.

156. M. Rasmusson, A. Routh and B. Vincent, *Langmuir*, 2004, **20**, 3536–3542.

157. A. Fernández-Nieves, J.S. van Duijneveldt, A. Fernández-Barbero, B. Vincent and F.J. de las Nieves, *Phys. Rev. E*, 2001, **64**, 015603; 1–10.

158. B.R. Monaghan and H.L. White, *J. Gen. Physiol.*, 1935, **19**, 719.

159. C. Bondy, *Trans. Faraday Soc.*, 1939, **35**, 1093.

160. A. Vrij, *Pure Appl. Chem.*, 1976, **48**, 47.

161. J.F. Joanny, L. Leibler and P.G. de Gennes, *J. Polym. Sci. (Phys.)*, 1979, **77**, 1073.

162. P.I. Feigin and D.H. Napper, *J. Colloid Interface Sci.*, 1980, **74**, 567; 1980, **75**, 525.

163. P.R. Sperry, *J. Colloid Interface Sci.*, 1981, **87**, 375.

164. S. Asakura and F. Oosawa, *J. Polym. Sci.*, 1958, **33**, 183.

165. C.L. Sieglaff, *J. Polym. Sci.*, 1959, **41**, 319.

166. G.J. Fleer, J.H.M.H. Scheutjens and B. Vincent, *Am. Chem. Soc. Symp. Ser.*, 1984, **240**, 281–298.

167. J. Clarke and B. Vincent, *J. Colloid Interface Sci.*, 1981, **82**, 208–216.

168. B. Vincent, J. Clarke and K.G. Barnett, *Colloids Surf.*, 1986, **17**, 51–65.

169. B. Vincent, J. Edwards, S. Emmett and A. Jones, *Colloids Surf.*, 1986, **17**, 261–281.

170. A. Jones and B. Vincent, *Colloids Surf.*, 1989, **42**, 113–138.

171. A.J. Milling, B. Vincent, S. Emmett and A. Jones, *Colloids Surf.*, 1991, **57**, 185–195.

172. A. Jones and B. Vincent, *Colloids Surf.*, 1989, **42**, 113–138.

173. P.G. de Gennes, *Macromolecules*, 1981, **14**, 1637; 1982, **19**, 492.

174. E. Donath, A. Krabi, G.C. Allan and B. Vincent, *Langmuir*, 1996, **12**, 3425–3430.

175. E. Donath, D. Walther, A. Krabi, G.C. Allan and B. Vincent, *Langmuir*, 1996, **12**, 6263–6269.
176. E. Donath, A. Krabi, M. Nirschl, V.M. Shilov, M.I. Zharkikh and B. Vincent, *J. Chem. Soc., Faraday Trans.*, 1997, **93**, 115–119.
177. A. Krabi, G.C. Allan, E. Donath and B. Vincent, *Colloids Surf.*, 1997, **122**, 33–42.
178. B. Vincent, *Colloids Surf.*, 1987, **24**, 269–282.
179. B. Vincent, J. Edwards, S. Emmett and R. Croot, *Colloids Surf.*, 1988, **31**, 267–298.
180. P. Jenkins and B. Vincent, *Langmuir*, 1996, **12**, 3107–3113.
181. G. Vliegenthart, J. S. van Duijneveldt and B. Vincent, *Faraday Discuss. R. Soc. Chem.*, 2003, **123**, 65–74.
182. N. Cawdery, A. Milling and B. Vincent, *Colloid Surf.*, 1994, **86**, 239–249.
183. A.J. Milling and B. Vincent, *J. Chem. Soc., Faraday Trans.*, 1997, **93**, 3179–3183.
184. R. Dunleavey-Routh and B. Vincent, *J. Colloid Interface Sci.*, 2007, **309**, 119–125.
185. B. Vincent, C.A. Young and Th.F. Tadros, *Faraday Discuss. Chem. Soc.*, 1978, **74**, 337–347.
186. B. Vincent, P.F. Luckham and F.A. Waite, *J. Colloid Interface Sci.*, 1980, **73**, 508–521.
187. W. van Megen, I. Snook and B. Vincent, *J. Colloid Interface Sci.*, 1983, **92**, 262–264.
188. S. Harley, D.W. Thompson and B. Vincent, *Colloids Surf.*, 1992, **62**, 163–176.
189. B. Vincent, M. Jafelicci, P.F. Luckham and Th.F. Tadros, *J. Chem. Soc., Faraday Trans. 1*, 1980, **76**, 674–683.
190. P.F. Luckham, B. Vincent, C.A. Hart and Th.F. Tadros, *Colloids Surf.*, 1980, **1**, 281–293.
191. P.F. Luckham, J. McMahon, Th.F. Tadros and B. Vincent, *Colloids Surf.*, 1983, **6**, 83–95.
192. P.F. Luckham, Th.F. Tadros and B. Vincent, *Colloids Surf.*, 1983, **6**, 101–118.
193. P.F. Luckham, Th.F. Tadros and B. Vincent, *Colloids Surf.*, 1983, **6**, 119–133.
194. N.J. Marston, B. Vincent and N.G. Wright, *Prog. Colloid Polym. Sci.*, 1998, **109**, 278–282.
195. D.R. Skuse, Th.F. Tadros and B. Vincent, *Colloids Surf.*, 1986, **17**, 343–360.
196. R. Davies, D.E. Graham and B. Vincent, *J. Colloid Interface Sci.*, 1987, **116**, 88–89.
197. R. Davies, D.E. Graham and B. Vincent, *J. Colloid Interface Sci.*, 1988, **126**, 616–621.
198. P. Starck and B. Vincent, *Langmuir*, 2006, **22**, 5294–5300.
199. A.F. Routh and B. Vincent, *J. Colloid Interface Sci.*, 2004, **273**, 435–441.

200. A. Fernández-Barbero, A. Loxley and B. Vincent, *Prog. Colloid Polym. Sci.*, 2000, **115**, 84–87.
201. A. Fernández-Barbero and B. Vincent, *Phys. Rev. E*, 2000, **63**, 1–7.
202. D.R.E. Snoswell, T.J. Rogers, A.M. Howe and B. Vincent, *Langmuir*, 2005, **21**, 11439–11445.
203. D.E. Snoswell, P. Creaton and B. Vincent, in preparation.
204. N. Marston and B. Vincent, *Langmuir*, 1997, **13**, 14–22.
205. K.T. Rees, PhD thesis, University of Bristol, 2005.
206. V. Nerapusri, J.L. Keddie, B. Vincent and I. Bushnak, *Langmuir*, 2006, **22**, 5036–5041.

Chapter 2

Synthesis of Poly(N-isopropylacrylamide) Microgel Particles Containing Gold Nanoshell Cores with Potential for Triggered De-swelling

Paul Luckham, Carlo Strazza, Pierre Bussierre, Paulo Nassari and Neil Patel

DEPARTMENT OF CHEMICAL ENGINEERING AND CHEMICAL TECHNOLOGY, IMPERIAL COLLEGE LONDON, PRINCE CONSORT ROAD, LONDON SW7 2AZ, UK

Abstract

This work describes the synthesis of gold nanoshells (silica core–gold shell) coated with a thermo-responsive and biocompatible poly(N-isopropylacrylamide) hydrogel layer. The negatively charged silica cores were functionalised with poly(diallyldimethylammonium chloride), making them cationic. Successively, small negatively charged gold colloid particles were adsorbed onto their surface, by electrostatic attraction, as nucleation sites; finally, further reduction (sonication assisted) of gold onto the silica involved the coalescence of the gold particles into a complete gold shell. Such particles show interesting photo-thermal properties in that when irradiated with near-infrared radiation they heat up.

The encapsulation of the gold nanoshells within the hydrogel over-layer was carried out via surfactant-free emulsion polymerisation of N-isopropylacrylamide. Reversible volume changes of the composite hydrogel nanoparticles were observed as a function of the temperature, with a lower critical solution temperature around 35 °C, similar to that of the pure N-isopropylacrylamide particles.

New Frontiers in Colloid Science: A Celebration of the Career of Brian Vincent
Edited by Simon Biggs, Terence Cosgrove and Peter Dowding

2.1 Introduction

In common with all the other authors of chapters in this volume, one of us (Paul Luckham) studied for his PhD with Professor Brian Vincent. The subject area was the hetero-flocculation of polystyrene latex particles by small particles of opposite charge.[1-6] Essentially large polystyrene latex particles (2 µm) were flocculated by small positive polystyrene particles (200 nm), the aim being to determine whether flocculation could be controlled by the addition of the oppositely charged particles. In turned out that it could but the concept has found few outlets in terms of applications. However, with the advent of the term "nanotechnology" this concept has been used to prepare core–shell particles containing gold which have interesting optical properties.[7-15] In this chapter we report the preparation of such particles and their use as a template onto which to grow *N*-isopropylacrylamide (NIPAM) layers which also have intriguing properties.

Metal nanoshell particles consist of a dielectric or semiconducting core coated with a nanometre-scale metallic shell. These nanoparticles manifest a strong optical resonance that is dependent on the relative thickness of the particle core and its metallic shell. By varying the core and shell thicknesses, this optical resonance can be placed virtually anywhere across the visible or infrared regions of the optical spectrum. These core–shell particles can serve as constituents in a new class of materials that are capable of uniquely controlling radiation in the visible and infrared spectral regions.[7,8,10,15]

Solid metallic nanoparticles are well known for their attractive optical properties: a strong optical resonance and an extremely large and fast nonlinear optical polarisability associated with their plasmon frequency. These optical properties are accounted for extremely well by classical electromagnetic theory Mie scattering.[16] Although the general Mie scattering solution for a spherical particle consisting of concentric layers has been known for decades, it was only recently theoretically established that a configuration consisting of a metallic shell and a dielectric core should result in a particle with a plasmon-derived optical resonance variable over large regions of the electromagnetic spectrum.

This sensitive dependence of the optical resonance frequency on the structure of metal nanoshells has been demonstrated by Halas and co-workers who indicated that silica–gold nanoshells with core–shell ratios of more than 20 would have optical resonances shifted into the near-infrared region.[7]

Nanoparticles of this type are perhaps uniquely suited as dopants for providing localised photo-thermal heating to a host material, particularly when compared with molecular chromophores. For high quantum yield molecular chromophores that relax primarily via radiative pathways, the excitation energy would be transmitted out of an optically transparent host material as light and therefore not participate in heat transfer to the host matrix. For molecular chromophores that relax non-radiatively, the relaxation is controlled by the long-lived microsecond to millisecond lifetime of the non-radiative triplet state of the molecule, which, in addition to weak coupling to the phonon density of states of the host material, places a severe limitation on the rate of energy transfer both into and out of each individual chromophore dopant. In contrast, metallic nanoshells have extremely large absorption cross sections, approximately one million times larger than a typical molecular chromophore, picosecond relaxation times and very low quantum efficiency, so that the vast majority of the absorbed photon energy will dissipate via electron–phonon interactions within the nanoparticle and with its local environment, *i.e.* as heat. This, coupled with the wavelength tunability of nanoshells, means that if nanoshells are irradiated with the correct wavelength radiation they will heat up. Interestingly body tissues are somewhat transparent to near-infrared radiation; thus, by tuning the wavelength of the nanoshell absorption to the near-infrared, it should be possible to heat up nanoshells within the body by irradiating the body with light of the correct wavelength. This been demonstrated by Halas and co-workers who showed that such particles injected into a tumour heated up when the tumour was irradiated with near-infrared light such that the cells were killed.[10,13]

The solubility of many polymers is also temperature dependent. A polymer that has received considerable attention in the last twenty years is poly (*N*-isopropylacrylamide) (polyNIPAM), which has a lower consulate solution temperature of 34 °C when dissolved in water. Pelton and co-workers[17–19] and Vincent and co-workers[20–29] have also made microgel particles of this polymer which shows a considerable swelling transition at this temperature, such that below around 35 °C the particles are swollen whilst above it the particles contract. The effect can be quite considerable with volume changes approaching 100-fold for some systems, giving rise to considerable rheological changes in the system. Such microgel particles are made through a simple dispersion polymerisation route analogous to the preparation of polystyrene latex. Many potential uses of these microgels have been proposed including controlled release, rheology modifiers, to form thin films, to act as microlenses, *etc.* It has been shown for example that the rate of doxorubicin[30–32] (a common chemotherapeutic drug used in cancer treatment) release from a thin film of isopropylacrylamide particles is greater at high temperatures than at low temperatures. Thus a means of triggering such a temperature change is advantageous. In this chapter we report a study where we have merged these two concepts together to produce a nanoshell isopropylacrylamide core–shell particle which undergoes changes when irradiated with near-infrared radiation.

2.2 Materials and Methods

The preparation of gold nanoshell–NIPAM hydrogels requires many stages which are detailed below. At each stage minor variations in the reaction conditions are possible to change principally the particle size. Here we present the method which was found to be the most reliable and appropriate for our purpose.

2.2.1 Preparation of Silica Particles

The silica particles were prepared using the method of Stober *et al.*[33] In brief, 4 ml of ammonia (30% NH$_3$ as NH$_4$OH assay ex Aldrich) was added to 250 ml of ethanol in a conical flask. The mixture was stirred vigorously for 2 minutes to ensure its homogeneity and then an aliquot (20 ml) of tetraethylorthosilane was added dropwise over a period of around 10 minutes. The dispersion was stirred for 20 hours at room temperature; initially the dispersion was transparent, but it became opalescent overnight. In order to remove excess reagents, the dispersion was dialysed in a bath of deionised water, for seven days, with at least three changes of the water per day.

The above procedure yielded particles with a radius of around 50 nm although different particle sizes could be prepared by changing the amount of ammonia in the ethanol.

2.2.2   Preparation of Positively Charged Silica Particles

In order to prepare gold nanoshells negatively charged gold particles are adsorbed onto positively charged silica and then the gold shell is grown around the silica. Thus it is necessary to prepare positively charged silica. In other work this has been achieved by reacting the silica with aminopropyltriethoxysilane.[34] We followed this procedure but found that the silica dispersions were not stable for long time periods; over the course of a week the silica flocculated and it was noted that this corresponded to a gradual decline in the zeta potential of the silica from +30 mV down to +5 mV or less. Therefore it was decided to functionalise the silica with a positively charged polyelectrolyte, poly (diallyldimethylammonium chloride) (PDADMAC). This proved to be much simpler and produced particles stable for many months. In brief, an excess of PDADMAC (100 µl of an 80% solution) was added to a 100 ml aliquot of the vigorously stirred silica nanoparticle solution, and the mixture was allowed to adsorb for 2 hours. The silica particles were then centrifuged and redispersed in water. The zeta potential of the silica particles was between +15 mV and +30 mV and was stable against flocculation for at least 6 months.

2.2.3   Preparation of Gold Nanoparticles

The preparation of small gold sols was achieved by following the method of Duff *et al.*[35,36] A vessel (e.g. a 100 ml round-bottomed flask) was charged, while stirring, with 45.5 ml of distilled water and then the following materials were added, in order:

   (i) 0.5 ml of a 1 M solution of sodium hydroxide (Aldrich);
  (ii) 1 ml of an aqueous solution containing the reducing agent THPC (ex Fluka) (1.2 ml of 80% aqueous solution diluted to 100 ml with water);
 (iii) 2 ml of a 25 mM dark-aged (3–4 days) stock tetrachloroauric acid ($HAuCl_4$) solution.

The reaction mixture was stirred for 5 minutes between the addition of THPC and that of tetrachloroauric acid. This method produced small gold particles (approximately 3 nm radius), with the sol appearing as a dark brown

Figure 2.1 Gold nanoparticles as prepared by the THPC route.

(cola) coloured solution (see Figure 2.1). A transmission electron microscopy (TEM) image of the gold nanoparticles is shown in Figure 2.2.

2.2.4 Attachment of Gold Nanoparticles to Silica Cores

The procedure simply involves the gentle mixing of a fixed concentration of the positively charged silica nanoparticles with various concentrations of the negatively charged gold colloid prepared. The concentration of adsorbed gold was determined via the depletion method where the gold-coated silica particles were removed by centrifugation and the concentration of any unadsorbed gold determined from the absorption at 500 nm; this method also gives us a pseudo-adsorption isotherm for the gold adsorption onto the silica. A TEM image of the resulting particles is shown in Figure 2.3.

2.2.5 Growth of Gold Nanoshells

Initially a 25 mM solution of $HAuCl_4$ in water was prepared and left to age in the dark for a week. It is not clear why this step is necessary, but if it is not taken poorer quality nanoshells were produced.

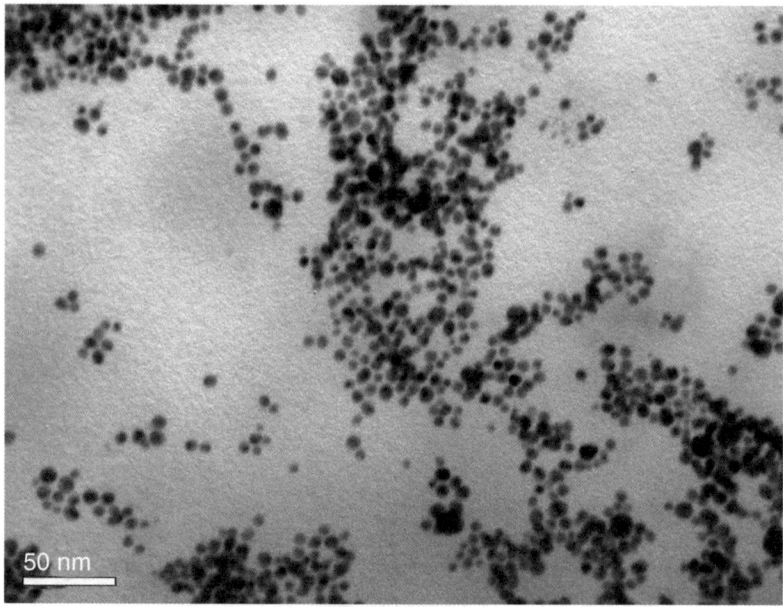

Figure 2.2   TEM image of the gold nanoparticles prepared by the THPC route.

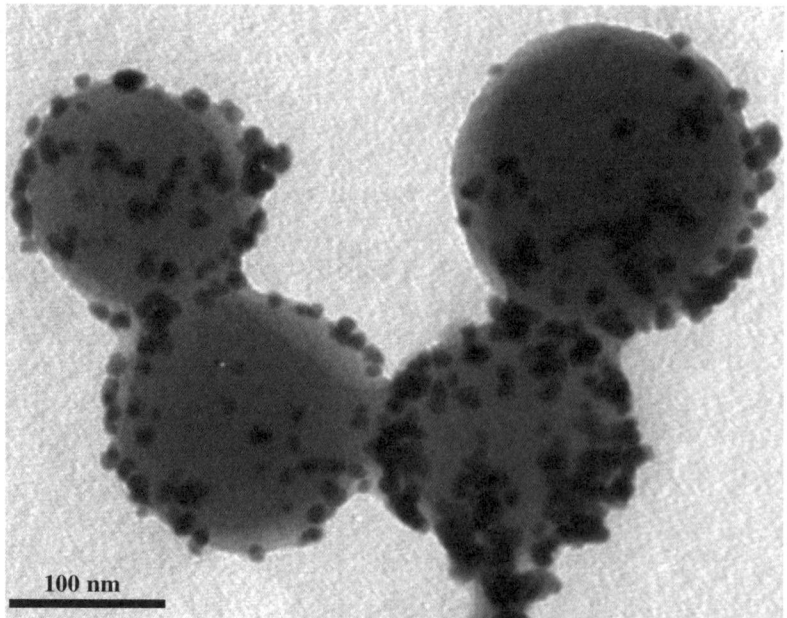

Figure 2.3   TEM image of the gold nanoparticles adsorbed on silica.

Figure 2.4 Example of the gold nanoshells produced during the work reported.

In a reaction flask, 100 mg of potassium carbonate (K_2CO_3) was dissolved in 400 ml of distilled water; then 6 ml of the aged $HAuCl_4$ solution was added and left to react overnight, with protection from light. At the beginning the solution appeared transparent yellow, and then slowly turned colourless.

An amount of 2 ml of the silica suspension bearing the adsorbed gold nanoparticles was injected to a vigorously stirred aliquot (40 ml) of the colourless solution of potassium carbonate and tetrachloroauric acid. An aliquot of formaldehyde (0.5 ml) was then injected; the reaction proceeded for 24 hours with protection from light. It was found that if the samples were sonicated in an ultrasonic bath during the first hour of the reaction, more complete nanoshells more stable colloidally were formed. The nanoshells were then centrifuged and redispersed in water and could be virtually any colour dependent on the size of the silica particles, the gold nanoparticles and the thickness of the gold film and varied from red through green to blue and then almost colourless; an example is shown in Figure 2.4. A TEM image of the resulting nanoshells is shown in Figure 2.5.

2.2.6 Synthesis of Hydrogel-coated Gold or Gold Nanoshells

An amount of 200 ml of either a gold dispersion or a gold nanoshell dispersion was placed into a three-necked round-bottomed flask (1 l) equipped with a reflux condenser and an inlet for nitrogen. The solution was vigorously stirred, and nitrogen was bubbled through it for 1 hour to completely remove any

Figure 2.5 TEM image of the gold nanoshells. Note that there appears to be two populations of silica-coated particles here, one where a relatively thick almost complete shell is formed and one (bottom right-hand corner of the image) where the gold particles have only grown a small amount.

oxygen, which can trap radicals and stop the polymerisation process. Sodium oleate solution (7.46 ml, 0.001 M) was then added to the gold nanoshell dispersion under nitrogen, and the mixture was stirred for another hour. NIPAM (1.376 g) was then added together with 100 mg of methylenebisacrylamide and stirred for 15 minutes to give homogeneity. The solution was then heated to 70 °C and then 0.084 g of ammonium persulfate, dissolved in 10 ml of water, was added to start the polymerisation. The mixture was allowed to react for 8 hours under a nitrogen atmosphere. The final mixture was centrifuged and the supernatant was carefully separated to remove any unreacted materials, soluble side products and oligomers. The purified nanoparticles were then diluted with distilled water and redispersed.

2.3   Results and Discussion

The synthesis of silica and gold particles is straightforward and well reported and a summary of the results has been presented in the experimental description above. Here we concentrate on the properties of the gold-coated silica particles and the nanoshell core–hydrogel shell particles.

2.3.1   Adsorption of Gold Nanoparticles onto Positively Charged Silica Particles

An adsorption isotherm for the adsorption of gold nanoparticles onto silica was obtained by adding known concentrations of gold to the silica and determining the concentration of gold remaining in solution after adsorption by measuring the absorbance of light at 500 nm using a UV–visible spectrometer. The results are presented in Figure 2.6. It can be seen that the adsorption is of high-affinity type, similar to a polymer adsorption isotherm or to the adsorption of small positive polystyrene particles onto large positive polystyrene particles. It can also be seen that the fractional coverage of the gold on the silica, θ, is quite low, around 0.22, which appears to be roughly in accord with the TEM image presented in Figure 2.3. This low coverage is likely to be due to the lateral electrical double layer repulsion of the adsorbed gold particles. It is worth noting that the TEM image appears somewhat patchy in terms of the adsorption; this may be due to drying effects in the preparation of the sample for microscopy. It must be recalled that the gold particles have a radius of only around 3 nm and so artefacts due to capillary effects induced by the drying process cannot be ruled out.

Figure 2.7 shows a comparison of the optical properties of the gold nanoparticles both in solution and when adsorbed onto the silica particles. It shows that when attached to silica the optical properties of the gold are similar although the intensity is much greater as the gold has been diluted by a factor of approximately 20 whilst the absorbance value only decreases by around 15%.

Plotted in Figure 2.8 are UV–visible spectra for four different samples: the gold nanoparticles, the gold particles adsorbed onto silica as in Figure 2.5 plus partially grown nanoshells and what we call fully grown nanoshells. It can be seen that as the gold layer grows the plasmon resonance peak shifts from around 530 nm to 650 nm with a considerable extension into the near-infrared region. If we look at the TEM image of the fully grown nanoshells we note that the shells are not actually fully grown, but are comprised of two types of

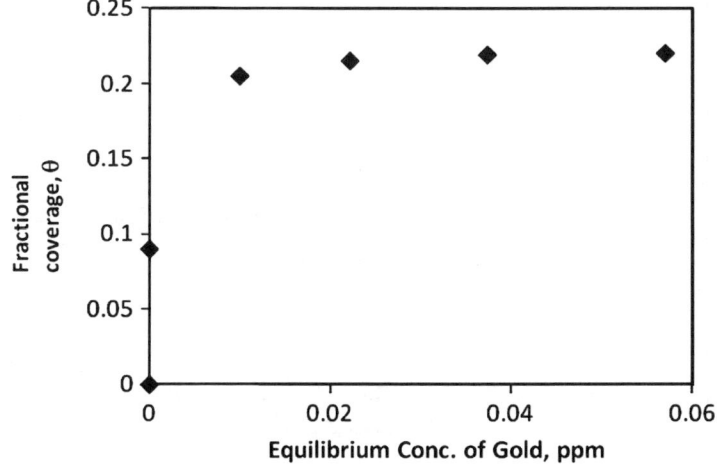

Figure 2.6   Adsorption isotherm for the adsorption of small negative gold particles (3–4 nm radius) onto large positive silica particles (60 nm radius).

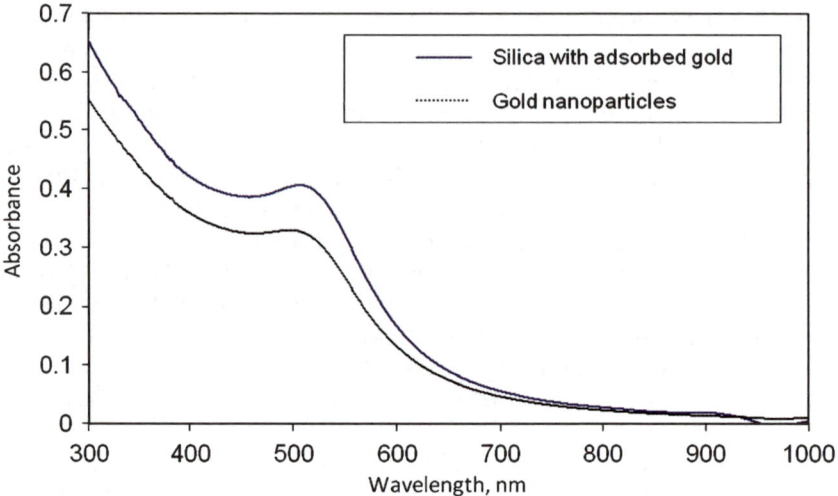

Figure 2.7   UV–visible spectra of the gold nanoparticles in solution and adsorbed onto positive silica particles.

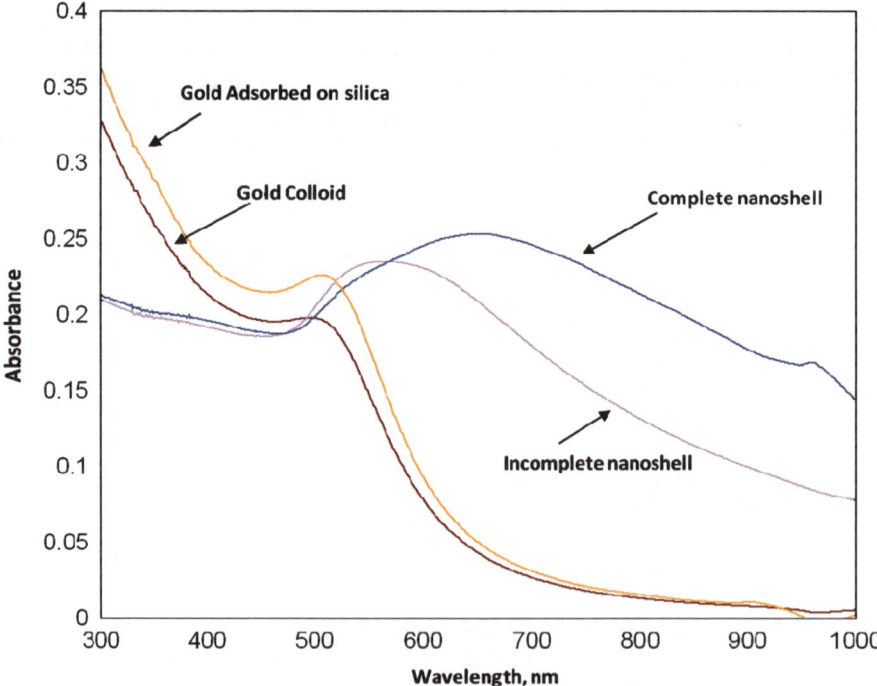

Figure 2.8   UV–visible spectra of the gold nanoparticles in solution, gold nanoparticles adsorbed onto positive silica particles, incomplete nanoshells and "complete" nanoshells.

particles: one set of particles which are almost fully grown although there are a few gaps; and some other particles where individual gold particles, admittedly larger that the original gold nanoparticles, can be clearly seen.

The next stage was to attempt to make a NIPAM hydrogel based around the gold nanoshells. In order to optimise the process initial attempts used 10 nm gold particles, as the process for making the nanoshells is fairly involved. Figure 2.9 presents a TEM image of such core–shell particles which was prepared by simply drying the particles at room temperature, *i.e.* below the lower consulate solution temperature of the NIPAM. It can be seen that there is a gold particle, roughly but not always in the centre of most of the particles. Some particles do not have a gold particle associated with them and occasionally some particles have two (not shown in the figure). At the time of writing we do not have any micrographs of the nanoshell–NIPAM particles.

In order to assess the thermal stability of the NIPAM particles and the gold core–NIPAM shell particles, the particle radii were measured from 20 to 60 °C using a Brookhaven ZetaPALS system. The results are presented in Figure 2.10. The most obvious feature of the results is that the effect of temperature on all the particles is very similar, namely that the particle radius decreases as the temperature increases by a factor of around 3 between 20 and 35 °C (which would correspond to an approximate 30-fold change in volume of the particle). The

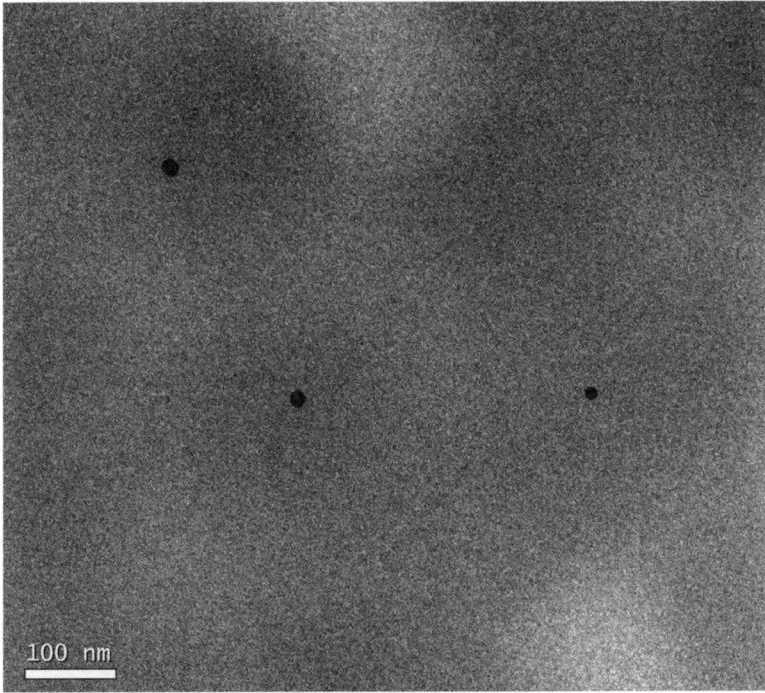

Figure 2.9 TEM image of polyNIPAM particles containing a gold nanoparticle core.

data are consistent with the well-known lower consulate solution temperature of polyNIPAM of 32 °C. Thus these composite particles have very similar properties to those of pure polyNIPAM particles. Secondly it is clear that the size of the nanoshell–NIPAM composite particles is roughly twice that of the NIPAM hydrogel, or the gold–NIPAM hydrogel; this is not surprising since the core of the particle is comprised of a gold nanoshell of radius of the order of 80 nm.

Finally we report some observations concerning the photo-thermal behaviour of these nanoshell–NIPAM composite particles. The nanoshell–NIPAM dispersion at room temperature is a translucent blue-black colour, whilst at 60 °C the sample is much more opaque due to the increase in refractive index of the particles on de-swelling. We note from Figure 2.8 that the nanoshell particles show a strong visible peak at around 650 nm. The laser of the Brookhaven instrument is a green laser with a wavelength of 532 nm, and, although there is some obscuration of the light of this wavelength in the UV–visible spectrometer, this is mostly due to scattering rather than absorption, and so irradiating the sample with this laser has no detectable effect on the particles, as seen in Figure 2.10. Thus it is not possible to unequivocally determine whether the nanoshell–NIPAM particles show any photo-thermal effects. However, red diode lasers such as a laser pointer have a wavelength of 635 nm. Thus we irradiated a cuvette of nanoshell–NIPAM particles with a laser pointer (1.5 mW) for 10 minutes at room temperature. The particles did not appear to change on illumination with the laser, but when the laser was turned off a thin faint cloudy line through the sample was observed, but only remained for a matter of seconds on turning the laser off. This suggests that the particles are exhibiting a photo-thermal effect and are heating up when irradiated by the laser and bringing about the phase transition of the NIPAM. On removing the laser the sample cools and swells, lowering the refractive index difference between the particles and solvent; also diffusion occurs, so the cloudy line disappears. We note that neither the pure NIPAM microgel nor the gold–NIPAM microgel exhibited this effect.

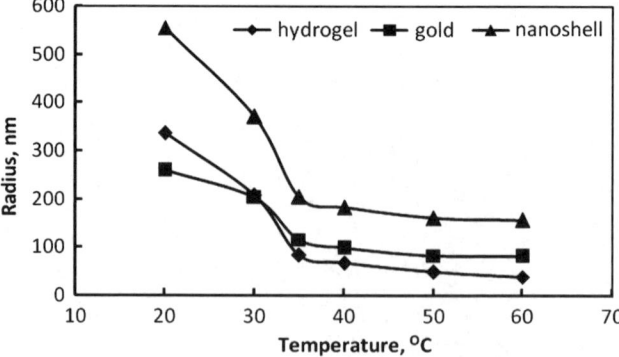

Figure 2.10   Particle size of the various NIPAM–gold particles prepared in this work as a function of temperature.

2.4 Concluding Remarks

We have demonstrated that it is possible to prepare polyNIPAM–gold nano-shell composite particles which have very similar properties to polyNIPAM microgel particles and give a strong indication that they heat up when irradiated with light of an appropriate wavelength. Such particles have potential as drug carriers where the swelling could be controlled by irradiation with near-infrared light. Such particles could also be used as valves in micro-fluidic devices.

Finally I would like to acknowledge that Brian Vincent has had the greatest influence on the philosophy of my research approach. I consider myself extremely fortunate that I was allocated a final year research project with Brian, although it was not my first choice! It was an easy choice to stay on for a PhD. The freedom I was given during my PhD to pursue my interests I greatly appreciated and certainly helped me subsequently. In those days Brian had a great deal of time and interest to spend with his students, although even then we were rather worried if he put his lab coat on! I recall with great fondness writing the first paper from my PhD in the Coronation Tap in Clifton, although I do not think we made much progress on the discussion! Finally as a person Brian does not seem to have changed much from those early days, which is meant as a compliment! I would be very pleased if people thought something similar of me in a few years time. My final hope is that Brian continues to have fun whilst pursuing his work, a trait that he has always shown ever since we first met.

References

1. P. Luckham, B. Vincent, C.A. Hart and T.F. Tadros, The controlled flocculation of particulate dispersions using small particles of opposite charge: 1. Sediment volumes and morphology, *Colloids Surf.*, 1980, **1**(3–4), 281–293.
2. B. Vincent, M. Jafelicci, P.F. Luckham and T.F. Tadros, Adsorption of small, positive particles onto large, negative particles in the presence of polymer: 2. Adsorption equilibrium and kinetics as a function of temperature, *J. Chem. Soc, Faraday Trans. 1*, 1980, **76**, 674–682.
3. B. Vincent, C.A. Young and T.F. Tadros, Adsorption of small, positive particles onto large, negative particles in the presence of polymer: 1. Adsorption isotherms, *J. Chem. Soc, Faraday Trans. 1*, 1980, **76**, 665–673.
4. P.F. Luckham, B. Vincent, J. McMahon and T.F. Tadros, The controlled flocculation of particulate dispersions using small particles of opposite charge: 2. Investigation of floc structure using a freeze-fracture technique, *Colloids Surf.*, 1983, **6**(1), 83–95.
5. P.F. Luckham, B. Vincent and T.F. Tadros, The controlled flocculation of particulate dispersions using small particles of opposite charge: 3. Investigation of floc structure using rheological techniques, *Colloids Surf.*, 1983, **6**(2), 101–118.

6. P.F. Luckham, B. Vincent and T.F. Tadros, The controlled flocculation of particulate dispersions using small particles of opposite charge: 4. Effect of surface coverage of adsorbed polymer on heteroflocculation, *Colloids Surf.*, 1983, **6**(2), 119–133.
7. S.J. Oldenburg, J.B. Jackson, S.L. Westcott and N.J. Halas, Infrared extinction properties of gold nanoshells, *Appl. Phys. Lett.*, 1999, **75**(19), 2897–2899.
8. S. Lal, S.L. Westcott, R.N. Taylor, J.B. Jackson, P. Nordlander and N.J. Halas, Light interaction between gold nanoshells plasmon resonance and planar optical waveguides, *J. Phys. Chem. B*, 2002, **106**(22), 5609–5612.
9. N. Halas, Plasmonic nanostructures and their applications in biosensing, *Abstr. Papers Am. Chem. Soc.*, 2003, **225**, U988.
10. L.R. Hirsch, R.J. Stafford, J.A. Bankson, S.R. Sershen, B. Rivera, R.E. Price, J.D. Hazle, N.J. Halas and J.L. West, Nanoshell-mediated near-infrared thermal therapy of tumors under magnetic resonance guidance, *Proc. Natl Acad. Sci. USA*, 2003, **100**(23), 13549–13554.
11. J.L. West, N. Halas and S.R. Sershen, Optically-responsive nanoshell composites, *Abstr. Papers Am. Chem. Soc.*, 2003, **225**, U522.
12. N. Halas, Nanoshells as multimodality nanoscale sensors, *Abstr. Papers Am. Chem. Soc.*, 2004, **227**, U107.
13. C. Loo, A. Lin, L. Hirsch, M.H. Lee, J. Barton, N. Halas, J. West and R. Drezek, Nanoshell-enabled photonics-based imaging and therapy of cancer, *Technol. Cancer Res. Treat.*, 2004, **3**(1), 33–40.
14. D.P. O'Neal, L.R. Hirsch, N.J. Halas, J.D. Payne and J.L. West, Photothermal tumor ablation in mice using near infrared-absorbing nanoparticles, *Cancer Lett.*, 2004, **209**(2), 171–176.
15. L.R. Hirsch, A.M. Gobin, A.R. Lowery, F. Tam, R.A. Drezek, N.J. Halas and J.L. West, Metal nanoshells, *Ann. Biomed. Eng.*, 2006, **34**(1), 15–22.
16. G. Mie, Articles on the optical characteristics of turbid tubes, especially colloidal metal solutions, *Annalen der Physik*, 1908, **25**, 377.
17. R.H. Pelton and P. Chibante, Preparation of aqueous lattices with N-isopropylacrylamide, *Colloids Surf.*, 1986, **20**(3), 247–256.
18. R.H. Pelton, H.M. Pelton, A. Morphesis and R.L. Rowell, Particle sizes and electrophoretic mobilities of poly(N-isopropylacrylamide) latex, *Langmuir*, 1989, **5**(3), 816–818.
19. W. McPhee, K.C. Tam and R. Pelton, Poly(N-isopropylacrylamide) latices prepared with sodium dodecyl-sulfate, *J. Colloid Interface Sci.*, 1993, **156**(1), 24–30.
20. M.J. Snowden, D. Thomas and B. Vincent, Use of colloidal microgels for the absorption of heavy-metal and other ions from aqueous-solution, *Analyst*, 1993, **118**(11), 1367–1369.
21. M.J. Snowden and B. Vincent, Flocculation of poly(*N*-isopropylacrylamide) latices in the presence of nonadsorbing polymer, *ACS Symp. Ser.*, 1993, **532**, 153–160.
22. M.J. Snowden and B. Vincent, Flocculation of microgels under Brownian and dynamic conditions, *Abstr. Papers Am. Chem. Soc.*, 1993, **205**, 137-COLL.

23. M.J. Snowden N.J. Marston and B. Vincent, The effect of surface modification on the stability characteristics of poly(N-isopropylacrylamide) latices under Brownian and flow conditions, *Colloid Polym. Sci.*, 1994, **272**(10), 1273–1280.

24. B.R. Saunders and B. Vincent, Thermal and osmotic deswelling of poly (NIPAM) microgel particles, *J. Chem. Soc., Faraday Trans.*, 1996, **92**(18), 3385–3389.

25. M.J. Snowden, B.Z. Chowdhry, B. Vincent and G.E. Morris, Colloidal copolymer microgels of *N*-isopropylacrylamide and acrylic acid: pH, ionic strength and temperature effects, *J. Chem. Soc., Faraday Trans.*, 1996, **92**(24), 5013–5016.

26. H.M. Crowther, B.R. Saunders, S.J. Mears, T. Cosgrove, B. Vincent, S.M. King and G.E. Yu, Poly(NIPAM) microgel particle de-swelling: a light scattering and small-angle neutron scattering study, *Colloids Surf. A*, 1999, **152**(3), 327–333.

27. P.J. Dowding, B. Vincent and E. Williams, Preparation and swelling properties of poly(NIPAM) "minigel" particles prepared by inverse suspension polymerization, *J. Colloid Interface Sci.*, 2000, **221**(2), 268–272.

28. M. Bradley and B. Vincent, Interaction of nonionic surfactants with copolymer microgel particles of NIPAM and acrylic acid, *Langmuir*, 2005, **21**(19), 8630–8634.

29. V. Nerapusri, J.L. Keddie, B. Vincent and I.A. Bushnak, Swelling and deswelling of adsorbed microgel monolayers triggered by changes in temperature, pH, and electrolyte concentration, *Langmuir*, 2006, **22**(11), 5036–5041.

30. V. Chytry and K. Ulbrich, Conjugate of doxorubicin with a thermosensitive polymer drug carrier, *J. Bioactive Compat. Polym.*, 2001, **16**(6), 427–440.

31. H. Doo and L.A. Lyon, The photophysics of doxorubicin in poly (*N*-isopropylacrylamide) microgels., *Abstr. Papers Am. Chem. Soc.*, 2006, **231**, COLL254.

32. M. Hruby, J. Kucka, O. Lebeda, H. Mackova, M. Babic, C. Konak, M. Studenovsky, A. Sikora, J. Kozempel and K. Ulbrich, New bioerodable thermoresponsive polymers for possible radiotherapeutic applications, *J. Controlled Release*, 2007, **119**(1), 25–33.

33. W. Stober, A. Fink and E. Bohn, Controlled growth of monodisperse silica spheres in micron size range, *J. Colloid Interface Sci.*, 1968, **26**(1), 62–69.

34. T. Pham, J.B. Jackson, N.J. Halas and T.R. Lee, Preparation and characterization of gold nanoshells coated with self-assembled monolayers, *Langmuir*, 2002, **18**(12), 4915–4920.

35. D.G. Duff, A. Baiker and P.P. Edwards, A new hydrosol of gold clusters: 1. Formation and particle-size variation, *Langmuir*, 1993, **9**(9), 2301–2309.

36. D.G. Duff, A. Baiker, I. Gameson and P.P. Edwards, A new hydrosol of gold clusters: 2. A comparison of some different measurement techniques, *Langmuir*, 1993, **9**(9), 2310–2317.

Chapter 3

Polymer Chemistry, Hypervelocity Physics and the Cassini Space Mission

Steven P. Armes

DEPARTMENT OF CHEMISTRY, DAINTON BUILDING,
UNIVERSITY OF SHEFFIELD, BROOK HILL, SHEFFIELD
S3 7HF, UK

Abstract

My PhD studies with Brian Vincent were concerned with the synthesis of electrically conductive latex particles based on polypyrrole and polyacetylene. I have continued to work in this fruitful field for the last twenty years (long after the funding dried up!) and in the last decade or so a rather esoteric application has been found for these fascinating particles. Due to their electrically conductive nature, polypyrrole-based particles can be accelerated up to hypervelocities ($>1\,\mathrm{km\,s^{-1}}$) using a high-voltage Van de Graaff accelerator. In collaboration with space physicists at the University of Kent, UK (and more recently also in Germany and Japan) we have designed a range of polypyrrole-based particles containing various elements (C, Br, Si, S, *etc.*). Such particles are excellent mimics for understanding the behaviour of micrometeorites (also known as 'cosmic dust'), which are typically travelling at hypervelocities through our solar system. Our results have implications for understanding some of the data being transmitted back to Earth by Cassini, an unmanned space probe that is currently orbiting Saturn and probing the chemical composition of the dust particles that comprise its rings. Thus this work typifies several aspects of Brian's long and illustrious career: fundamental science, interesting and unexpected applications, informal (and international) collaborations with scientists in other disciplines and, most importantly, it has the great merit of being fun!

New Frontiers in Colloid Science: A Celebration of the Career of Brian Vincent
Edited by Simon Biggs, Terence Cosgrove and Peter Dowding
© The Royal Society of Chemistry 2008

3.1 Introduction

After graduating with a BSc degree at Bristol in 1983, my PhD topic with Brian Vincent was concerned with the synthesis of conducting polymer latex particles based on polypyrrole and polyacetylene. After finally leaving Bristol in 1987, I continued to work in this area for the next twenty-odd years, so it has certainly had a major influence on my research career! In the last decade or so, I began collaborating informally with a space physicist, Professor Mark Burchell, at the University of Kent, UK, to investigate a rather esoteric application for conducting polymer-based particles. An overview of this informal collaboration is given in this chapter.

3.2 Organic Conducting Polymers

Organic conducting polymers became widely recognised as a new class of materials in 1977 after publication of a seminal *Chemical Communications* article by Shirakawa *et al.*[1] Classic examples include polyacetylene, polypyrrole and polyaniline; certain other conducting polymers are the active components in the plastic light-emitting diodes developed by Holmes and co-workers[2] at Cambridge that led to the highly successful spin-out company Cambridge Display Technology (CDT). This chapter is concerned with colloidal particles based on polypyrrole (PPy).

3.3 Synthesis and Properties of Polypyrrole

Pyrrole can be readily polymerised in acidic aqueous solution at ambient temperature using common chemical oxidants such as $FeCl_3$ or $(NH_4)_2S_2O_8$.[3-5] In the absence of any colloidal stabiliser, PPy is obtained as a black precipitate (see Figure 3.1).

PPy is lightly cross-linked and hence insoluble in all known solvents. On average, there is one cationic charge for every 3–4 polymerised pyrrole units. This charge is delocalised along the highly conjugated PPy backbone, giving rise to a relatively high electrical conductivity of around $1-10\,S\,cm^{-1}$ at room

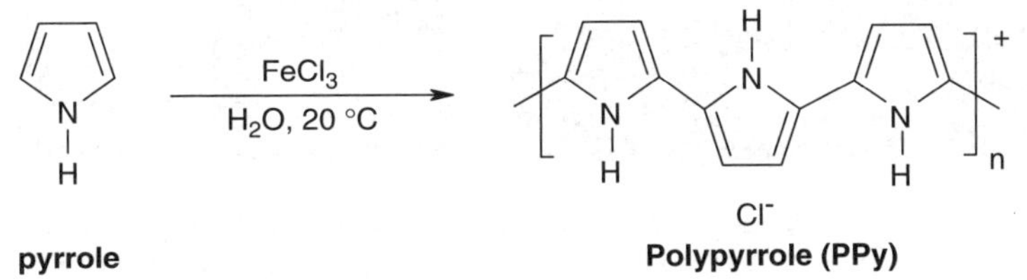

pyrrole   Polypyrrole (PPy)

Figure 3.1   Chemical synthesis of polypyrrole in aqueous solution and its idealised structure. Note the highly conjugated backbone, which is essential for efficient charge transport and high electrical conductivity.

temperature. This is comparable to the conductivity of a high-quality carbon black. Unlike polyacetylene, which suffers from chemical instability and rapid conductivity decay on long-term storage, the conductivity of PPy remains reasonably constant over time scales of months to years. Thus, by belatedly focusing my attention on PPy, I was able to abandon all the tedious and time-consuming air-sensitive techniques and protocols that I had developed for handling polyacetylene and hence ensure that I finished my PhD studies within a reasonable time scale. By the end of my PhD at Bristol, I had developed a facile aqueous dispersion polymerisation route to near-monodisperse, sterically stabilised PPy particles of around 100 nm in diameter.[6,7] Two common water-soluble polymers, poly(vinyl alcohol) and poly(N-vinylpyrrolidone), proved to be particularly effective polymeric stabilisers in such syntheses; in the absence of any stabiliser, PPy was invariably obtained as an amorphous precipitate with a globular morphology (see Figure 3.2). Originally, Brian and I had wanted to prepare colloidal forms of polypyrrole in order to overcome the problem of its notoriously poor processability, but the nonfilm-forming nature of these PPy particles meant that this approach eventually had to be abandoned.

As a young lecturer at Sussex University in the mid-1990s, I collaborated with the Defence Research Agency on the design of 'smart smoke bombs' for military camouflage applications. As part of this project, we exploited an approach originally pioneered by Wiersma et al.[8] of DSM Research to coat micrometre-sized, sterically stabilised polystyrene (PS) latexes with an ultrathin overlayer of electrically conductive PPy (see Figure 3.3).[9–12]

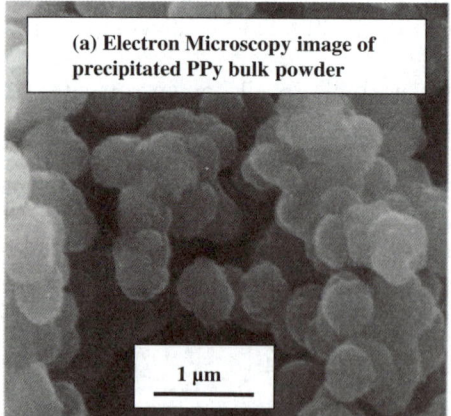
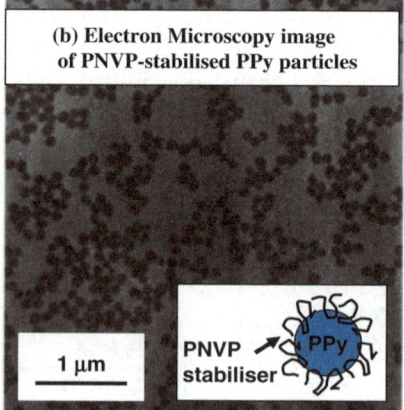

(a) Electron Microscopy image of precipitated PPy bulk powder

1 μm

(b) Electron Microscopy image of PNVP-stabilised PPy particles

1 μm

PNVP stabiliser   PPy

Figure 3.2 (a) Scanning electron micrograph of polypyrrole bulk powder prepared by precipitation using FeCl₃ oxidant; (b) transmission electron micrograph of sterically stabilised polypyrrole particles of approximately 100 nm diameter prepared under the same conditions using a suitable water-soluble polymer as a steric stabiliser. In this case the steric stabiliser is poly(N-vinylpyrrolidone) (PNVP).

(a)

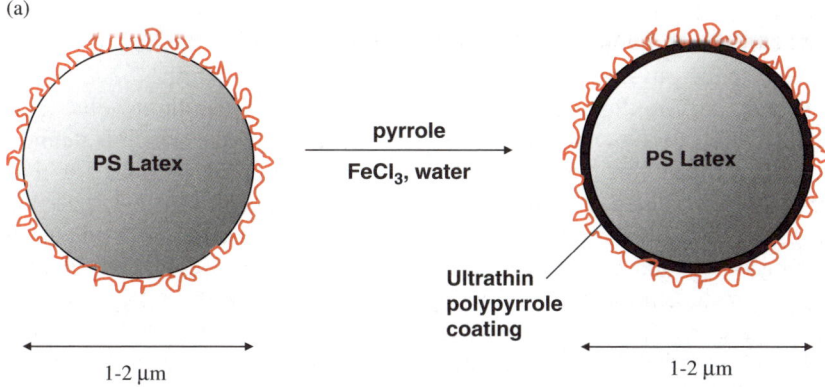

(b)

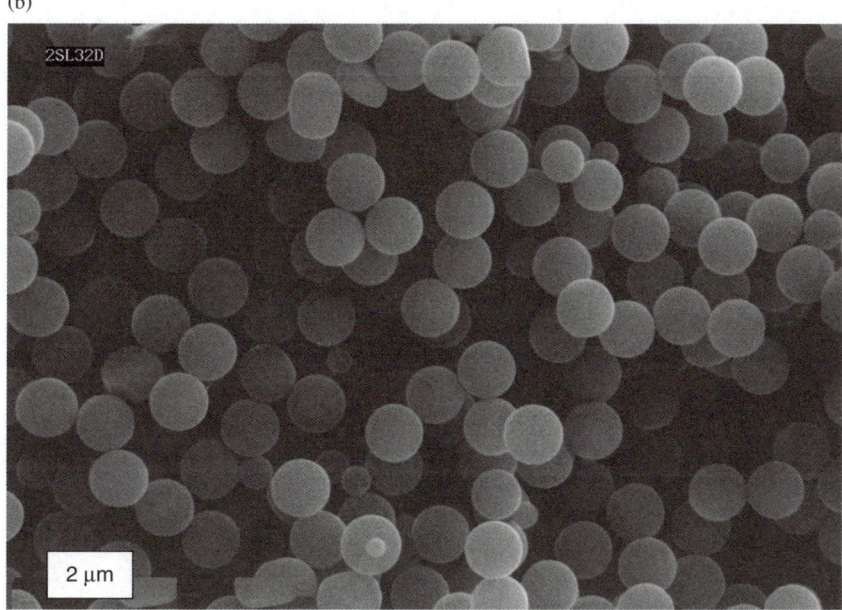

Figure 3.3 (a) Schematic of the synthesis of polypyrrole-coated polystyrene latex and (b) a representative scanning electron micrograph of polypyrrole-coated polystyrene latex (6.0 wt% polypyrrole loading by mass). Note the very smooth featureless morphology obtained for the polypyrrole overlayer.

3.4 Hypervelocity Acceleration Experiments

The military camouflage idea never really worked well enough to be developed further, but we returned to these PPy-coated latexes when we began to collaborate with Professor Mark Burchell at the University of Kent in 1997. Since the PPy coating was located on the outside of the electrically insulating PS latex, this 'core–shell' morphology allowed a relatively high surface charge to be accumulated, which in turn led to efficient acceleration of these PPy-coated PS particles using the high-field Van de Graaff instrument located in Professor Burchell's space physics laboratory at the University of Kent.[13]

This Van de Graaff instrument operates at 1.5 to 2.0 million volts. This is sufficiently high to allow access to the hypervelocity regime, which is defined as $>1\,\mathrm{km\,s^{-1}}$. Hypervelocities of $1-15\,\mathrm{km\,s^{-1}}$ were routinely obtained for micrometre-sized PPy-coated PS particles, while submicrometre-sized PPy-based particles could be accelerated to even higher hypervelocities.[7,14,15] Projectiles impinging on a solid target at such enormous speeds possess sufficient kinetic energy to produce impact craters many times larger than their original particle diameter (see Figure 3.4). Moreover, the impinging particles are generally completely volatilised to produce ionic plasma comprising charged atoms and molecules. Thus the chemical compositions of such plasma (and hence the original projectiles prior to their impact) can be readily deduced by time-of-flight mass spectroscopy analysis.[16] This is the key detection principle used by one of the instruments onboard the Cassini spacecraft (as discussed later in the chapter).

In the early 1990s my first PhD student, Mike Gill, serendipitously discovered the synthesis of polyaniline–silica nanocomposite particles during an ICI-sponsored CASE project.[17,18] Later we extended this approach to include polypyrrole–silica nanocomposite particles[19,20] and more recently to vinyl polymer–silica nanocomposite particles.[21,22] This synthesis involves an *in situ* polymerisation in the presence of an ultrafine aqueous silica sol (see Figure 3.5) and can be viewed as an example of controlled heteroflocculation, which has also been a fruitful research area for Brian over the last two decades.[23–25] Despite having silica-rich surfaces as judged by X-ray photoelectron spectroscopy (XPS),[20] the polypyrrole–silica nanocomposite particles can

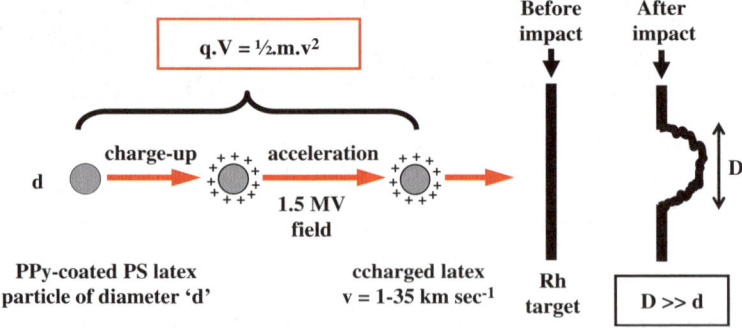

Figure 3.4   Schematic of a hypervelocity experiment. The impinging projectile is a conducting polymer-coated polystyrene latex particle, which can acquire surface charge and hence be accelerated up to hypervelocities using a high-voltage Van de Graaff instrument. The conducting polymer overlayer is essential: uncoated polystyrene latex particles cannot be accelerated since they are electrical insulators.

(a)

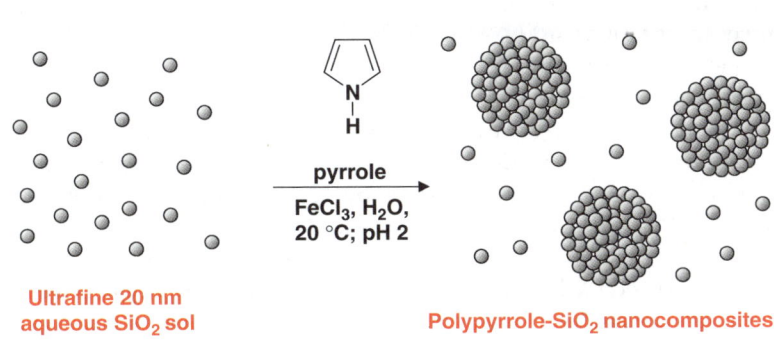

Ultrafine 20 nm aqueous SiO₂ sol   **Polypyrrole-SiO₂ nanocomposites**

(b)

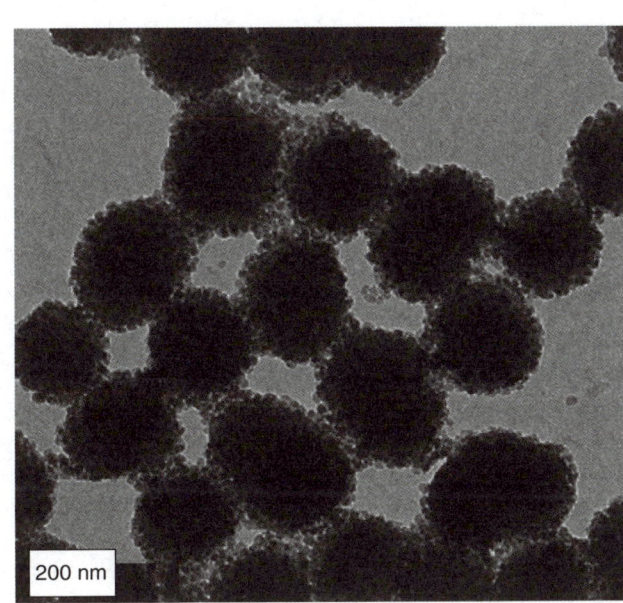

Figure 3.5   (a) Schematic of the formation of polypyrrole–silica nanocomposite particles obtained from the *in situ* polymerisation of pyrrole in the presence of an ultrafine aqueous silica sol; (b) typical transmission electron micrograph obtained for such polypyrrole–silica nanocomposite particles.

nevertheless be accelerated up to hypervelocities and these projectiles currently represent the best available mimics for silicate-based micrometeorites.[14]

3.5   Do Plasma Mass Spectra Originate from the Latex Core or the Shell?

The micrometre-sized PPy-coated PS latex projectiles comprise over 90% PS by mass.[9,11] However, although the PPy is the minor component, it is located on the outside of the projectile. Presumably this is why Kissel and Krueger[26] attempted to interpret and assign their plasma mass spectra (obtained from hypervelocity experiments using conducting polymer-coated PS latex particles originating from my research group) in terms of fragmentation of the conducting polymer coating, rather than the underlying PS latex core. However, Professor Burchell has examined the same PS latex coated with different conducting polymers, namely polypyrrole, polyaniline and poly(3,4-ethylenedioxythiophene).[11,27,28] In each case very similar mass spectra were obtained, which suggests that the ionic plasma generated during a hypervelocity impact is mainly due to the PS latex, rather than the conducting polymer shell.[29]

3.6   Kinetic Energy of a Hypervelocity Impact

A simple calculation using the well-known equation, kinetic energy (KE) = $(1/2)mv^2$, indicates that the KE for a micrometre-sized PPy-coated PS latex particle travelling at just $3\,km\,s^{-1}$ is approximately $470\,kJ\,mol^{-1}$. Thus there is sufficient KE available to break chemical bonds during a hypervelocity impact. However, this analysis assumes that all the KE of the impinging projectile is simply converted into chemical energy, whereas in fact some of this KE is also converted into light, heat and sound on impact, as well as considerable localised destruction of the metal target itself. Nevertheless, cleavage of chemical bonds does occur and some selectivity has been observed for low-KE impacts. For example, PS-based projectiles contain both aliphatic and aromatic bonds and it is observed that the weaker C–C bonds are broken in preference to the stronger C = C bonds at lower hypervelocities. Thus PPy-coated PS latex projectiles impinging under such conditions usually produce a strong signature at 91 amu, which is most likely to be due to the well-known tropylium cation.[16] To verify this tentative assignment, we also synthesised PPy-coated poly(4-bromostyrene) latex particles. Hypervelocity impact experiments using these projectiles led to the observation of a 1 : 1 doublet at 169 and 171 amu in the corresponding ionic plasma mass spectrum.[29] Since bromine has two isotopes, ^{79}Br and ^{81}Br, of approximately equal natural abundance, this spectral feature is clearly due to bromotropylium cation, thus supporting the original assignment of tropylium cation for the PS latexes. Finally, increasing the hypervelocity from $3\,km\,s^{-1}$ up to $10\,km\,s^{-1}$ leads to more than an order of magnitude increase in the KE for impinging PPy-coated PS latex projectiles. Under these more energetic conditions, there is sufficient excess KE to ensure that essentially all chemical bonds within the projectile are broken. In summary, various *molecular* ions are usually observed for PPy-coated PS projectiles impinging at lower hypervelocities, while *atomic* ions typically predominate at higher hypervelocities (see Figure 3.6).

3.7   The Cassini Space Mission and the CDA Detector

Cassini was launched in October 1997. It weighs nearly 6 tonnes, cost approximately $3.3 billion and is the largest unmanned spacecraft ever launched.

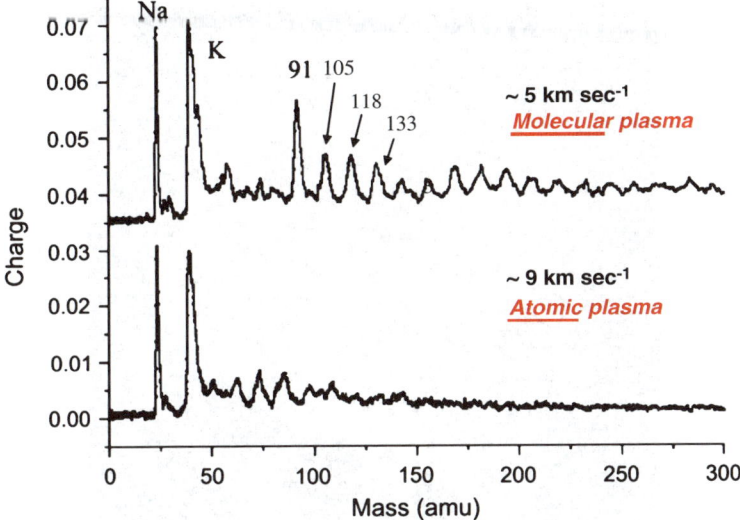

Figure 3.6 Time-of-flight mass spectra obtained for the cationic plasma generated after the hypervelocity impact of polypyrrole-coated polystyrene latex particles impinging at two velocities (as indicated) on a metal target.

Cassini reached Saturn in July 2004 and is now in orbit within its ring structure. One of its twelve scientific instruments is a 'cosmic dust' analyser (CDA; see Figure 3.7), which essentially comprises a metal target and a simple time-of-flight mass spectrometer.

As Cassini orbits Saturn, this CDA is continually bombarded by micrometeorites that are travelling at hypervelocities.[30] Total volatilisation of each micrometeorite occurs under these conditions, resulting in the generation of ionic plasma, which is then analysed by the mass spectrometer. The resulting plasma mass spectra are transmitted back to Earth, where space scientists try to determine the elemental compositions of the original impinging micrometeorites.

3.8 Four Classes of Micrometeorites

There are four main classes of micrometeorites that are typically found in outer space. These are (i) metallic particles (usually based on either iron or nickel), (ii) organic/carbonaceous particles, (iii) silicate-based particles and (iv) ice particles. NASA has utilised iron particles for Earth-based Van de Graaff accelerator experiments for more than four decades, primarily because these particles are readily available. Despite their relatively high polydispersity (see Figure 3.8), such metallic particles have high conductivities and can therefore easily acquire a high surface charge, which is a prerequisite for Van de Graaff-based hypervelocity experiments. However, the other three classes of micrometeorites are all electrical insulators; thus these materials cannot be accelerated.

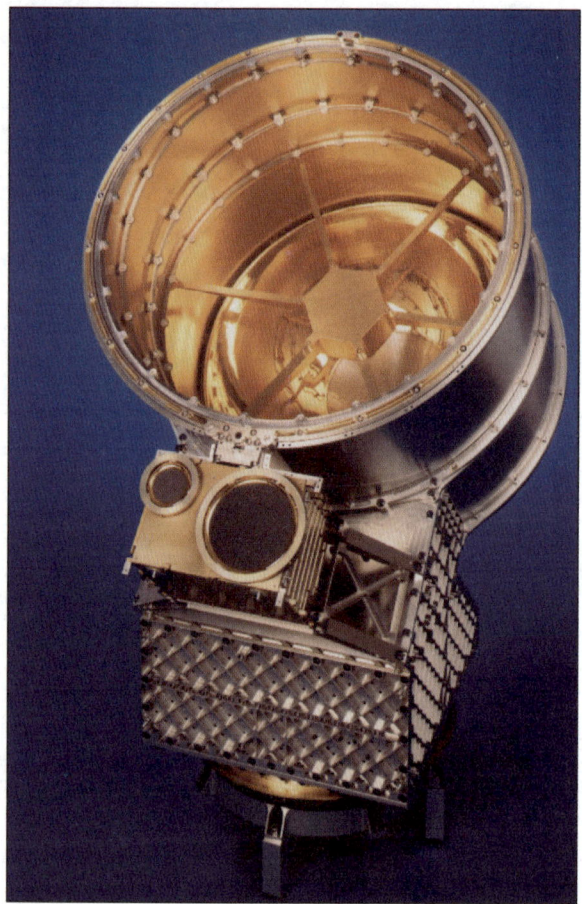

Figure 3.7 The cosmic dust detector (CDA) instrument, one of twelve instruments on the Cassini spacecraft, which began orbiting Saturn in July 2004. The CDA analyses the elemental compositions of micrometeorites (cosmic dust) that are converted into ionic plasma after impacting its rhodium target at hypervelocities. There are two duplicate CDA detectors on Earth for calibration purposes, one is in the UK and the other is in Heidelberg, Germany.

The typical carbon contents of our PPy-coated PS latexes are around 85% by mass; thus these well-defined core–shell latexes became the first *synthetic* projectiles to be used for mimicking the behaviour of organic/carbonaceous micrometeorites. It is worth emphasising that control experiments with uncoated charge-stabilised PS latexes were unsuccessful: no acceleration could be achieved with these particles, presumably because their surface charge density was too low. This confirms that the PPy overlayer is essential, since it allows the latex particles to acquire substantial surface charge and hence be accelerated using the Van de Graaff instrument (see Figure 3.4).

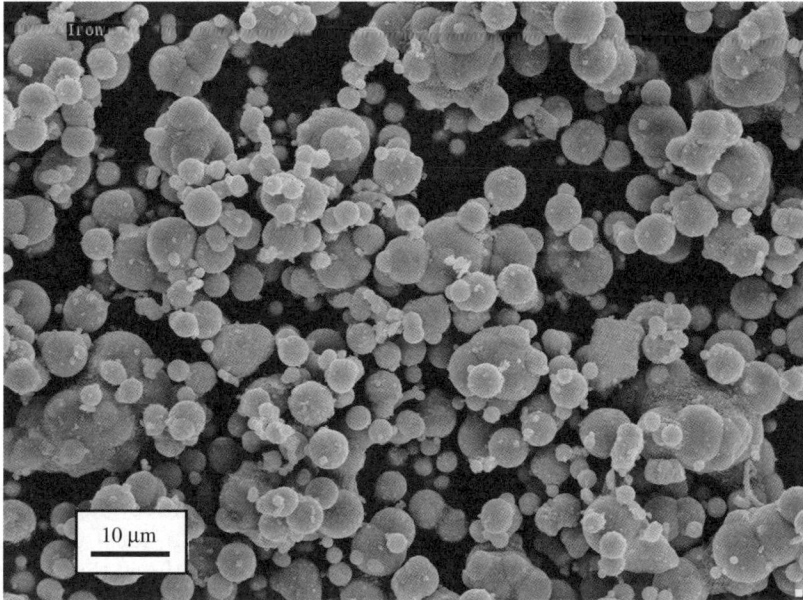

Figure 3.8   Scanning electron micrograph of the iron particles typically used as mimics for understanding the behaviour of metallic micrometeorites. Note the highly polydisperse nature of these 'model' particles compared to the near-monodisperse PPy-coated PS latex particles (see Figure 3.3).

Similarly, the PPy–SiO$_2$ nanocomposite particles proved to be reasonably good mimics for silicate-based micrometeorites.[14] Very recently, hypervelocity experiments conducted by Dr R. Srama in Heidelberg have confirmed that a silicon signal can be observed at 28 amu in ionic plasmas produced by such projectiles after their acceleration up to 20–30 km s^{-1}. Unfortunately, there are still no good mimics available for ice-based micrometeorites.

3.9 What is the Nature of the Volcanic Activity on Io?

Io is the third largest of Jupiter's moons. As it orbits Jupiter, Io is periodically squeezed by Jupiter's enormous gravitational field, which causes huge internal pressures within Io's crust. This is believed to account for Io's volcanic activity: apart from the Earth, Io is the only celestial body in our solar system that is known to be volcanically active. This activity was first observed during a Voyager mission in the 1970s and confirmed by a later Galileo mission. However, there is still some debate as to the nature of the volcanic plumes that are ejected hundreds of miles into space from Io's surface. The most likely explanation is that the ejecta comprise sulfur-based micrometeorites.[31] During its fly-by of Jupiter en route to Saturn, Cassini flew close to Io and attempted to detect sulfur-based micrometeorites in its vicinity.

3.10   Latex Mimics for Sulfur-rich Micrometeorites

The space physics community's great interest in Io led to a request for synthetic mimics of sulfur-rich micrometeorites. We took up this challenge and successfully prepared sulfur-rich latex particles using a bifunctional sulfur-based monomer, bis(4-vinylthiophenyl)sulfide (known as 'MPV'). These particles were rather polydisperse, but could be successfully coated with a thin overlayer of PPy to produce composite particles that contained 28% sulfur by mass (see Figure 3.9).[32]

XPS and Raman studies confirmed that core–shell morphologies were obtained, but the electrical conductivity of these PPy-coated latex particles was relatively low at $6 \times 10^{-5}\,S\,cm^{-1}$. Nevertheless, the particles were successfully accelerated up to 20–25 km s^{-1} using a Van de Graaff facility in Heidelberg and the ionic plasmas generated from impacts on a metal target confirmed that both S^{+} and S^{-} species were detected at 32 amu (see Figure 3.10).

Thus these terrestrial experiments confirmed that charged sulfur species are indeed present in ionic plasma generated during hypervelocity impacts of sulfur-rich projectiles. This positive result adds considerable confidence to the rather tenuous arguments used in interpreting the less-than-ideal data transmitted back to Earth by Cassini.[33] In summary, the hypothesis that Io's volcanic plumes are likely to be sulfur-rich[31] has been given additional credence, without being confirmed unequivocally.

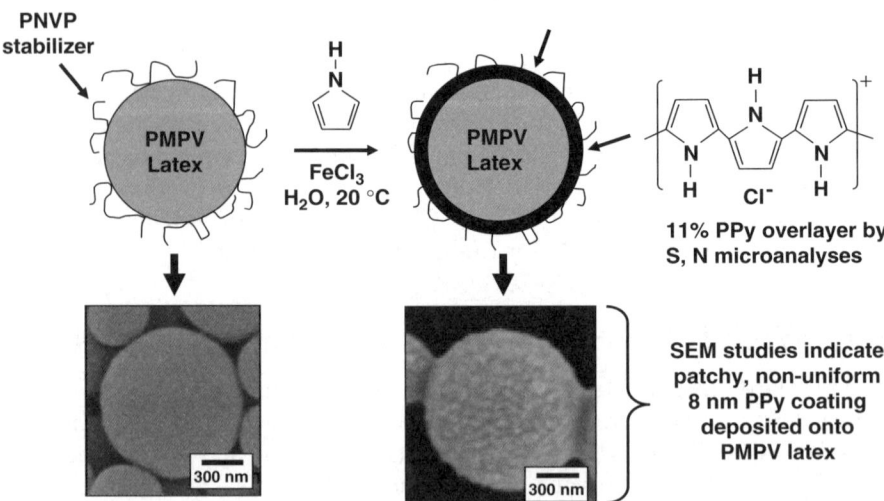

Figure 3.9   Schematic of the formation of polypyrrole-coated sulfur-rich latex particles obtained by depositing polypyrrole onto a sulfur-rich PMPV latex precursor from aqueous solution. The two high-magnification scanning electron micrographs were obtained before and after coating the sulfur-rich particles with polypyrrole. Note that a polypyrrole overlayer thickness of approximately 8 nm has discernible surface roughness compared to the relatively smooth uncoated PMPV latex.

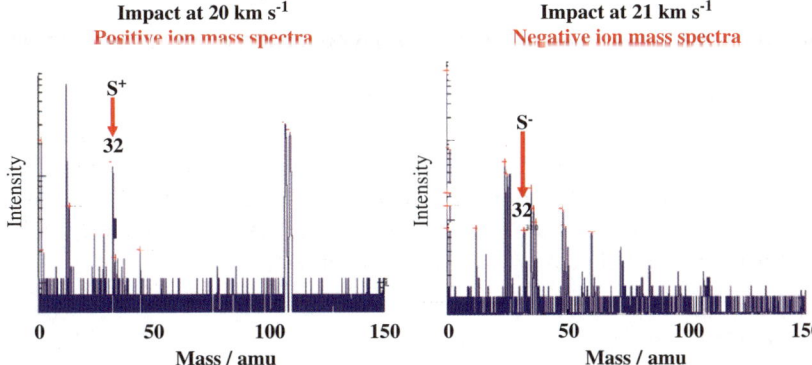

Figure 3.10 Time-of-flight mass spectra obtained for the cationic plasma (left) and the anionic plasma (right) generated after the hypervelocity impact of polypyrrole-coated sulfur-rich latex particles impinging at more than $20 \, km \, s^{-1}$ onto a metal target. Note that a signal at 32 amu (corresponding to either S^+ cation or S^- anion, respectively) is observed in each case. This confirms that these sulfur-rich organic latexes are useful mimics for understanding the behaviour of the sulfur-based micrometeorites that are postulated to originate from active volcanoes on Io.

3.11 Conclusions

In view of their relatively narrow size distributions, low particle densities and relative ease of synthesis, polypyrrole-based particles are model projectiles for laboratory experiments aimed at understanding the behaviour of micrometeorites. A wide range of projectiles have been designed that are useful mimics for understanding the behaviour of organic/carbonaceous micrometeorites, silicate-based micrometeorites and sulfur-rich micrometeorites. The results from these terrestrial hypervelocity experiments are expected to be useful for interpreting the mass spectra currently being transmitted back to Earth by Cassini, which is analysing the composition of cosmic dust particles within Saturn's rings.[24] This informal collaboration has certainly been enormous fun and it has also been converted into a 'schools' lecture that I have given to sixth form students at colleges in Brighton, Oxford, Croydon and Sheffield. Thus it incorporates some of the important aspects of Brian's illustrious academic career: basic research, a commitment to teaching and interdisciplinary (and international) collaborations.

Acknowledgements

I wish to thank Brian Vincent for inspiring me to enter the world of colloid/ polymer science and particularly for keeping me 'on track' during the darker days of my PhD studies. His friendship and support have also been invaluable in my later academic career. I also acknowledge my former PhD students and postdoctoral workers (Dr S. F. Lascelles, Dr D. B. Cairns, Dr M. A. Khan, Dr C. Barthet, Dr M. J. Percy and Dr S. Fujii) who all agreed to 'moonlight' in this area of space science applications for conducting polymer particles. I also thank two of my current PhD students, Damien Dupin and Andreas Schmid, for the TEM image shown in Figure 3.5b. It has been a particular pleasure to work with Prof. Mark Burchell at the University of Kent and also Dr Ralf Srama at the Max-Planck Institut für Kernphysik in Heidelberg. Finally, I acknowledge receipt of a Royal Society-Wolfson Research Merit Award, which has allowed me to buy the house that my wife Alison deserves.

References

1. H. Shirakawa, E.J. Louis, A.G. MacDiarmid, C.K. Chiang and A.J. Heeger, *Chem. Commun.*, 1977, 578.
2. J.H. Burroughes, D.D.C. Bradley, A.R. Brown, R.N. Marks, K. Mackay, R.H. Friend, P.L. Burns and A.B. Holmes, *Nature*, 1990, **347**, 539.
3. (a) A. Angeli and G. Cusmano, *Gazz. Chim. Ital.*, 1917, **47**, 207; (b) P. Pratezi, *Gazz. Chim. Ital.*, 1936, **66**, 215; (c) F. Hautiere-Cristofini, D. Kuffer and L.-T. Yu, *C. R. Acad. Sci. Ser. C*, 1973, **277**, 1323.
4. (a) A. Pron, Z. Kucharski, C. Budrowski, M. Zagorska, S. Krichene, J. Suwalski, G. Dehe and S. Lefrant, *J. Chem. Phys.*, 1985, **83**, 5923; (b) R.E. Myers, *J. Electron. Mater.*, 1986, **2**, 61.
5. S.P. Armes, *Synth. Met.*, 1987, **20**, 367.
6. (a) S.P. Armes and B. Vincent, *J. Chem. Soc., Chem. Commun.*, 1987, 288; (b) S.P. Armes, J.F. Miller and B. Vincent, *J. Colloid Interface Sci.*, 1987, **118**, 410; (c) S.P. Armes, M. Aldissi, G.C. Idzorek, P.W. Keaton, L.J. Rowton, G.L. Stradling, M.T. Collopy and D.B. McColl, *J. Colloid Interface Sci.*, 1991, **141**, 119.
7. Many years later, these 100 nm PPy particles proved to have the highest hypervelocities (around 35 km s^{-1} = 70 000 mph = Mach 100) of a wide range of conducting polymer-based projectiles evaluated by our space physicist collaborators.
8. (a) A.E. Wiersma, L.M.A. vd Steeg and T.J.M. Jongeling, *Synth. Met.*, 1995, **71**, 2269; (b) A.E. WiersmaL.M.A. vd Steeg *Eur. Pat. Appl.* (1994) EP589529; (c) L.G.B. Bremer, M.W.C.G. Verbong, M.A.M. Webers and M.A.M.M. van Doorn, *Synth. Met.*, 1997, **84**, 355.
9. S.F. Lascelles and S.P. Armes, *Adv. Mater.*, 1995, **7**, 864.
10. C. Perruchot, M.M. Chehimi, M. Delamar, S.F. Lascelles and S.P. Armes, *Langmuir*, 1996, **12**, 3245.
11. S.F. Lascelles and S.P. Armes, *J. Mater. Chem.*, 1997, **7**, 1339.
12. S.F. Lascelles, S.P. Armes, S.Y. Luk, P. Zhdan, A.M. Brown, S.J. Greaves, S.R. Leadley and J.F. Watts, *J. Mater. Chem.*, 1997, **7**, 1349.
13. M.J. Burchell, M.J. Cole, S.F. Lascelles, M.A. Khan, C. Barthet, S.A. Wilson, D.B. Cairns and S.P. Armes, *J. Phys. D: Appl. Phys.*, 1999, **32**, 1719.
14. M.J. Burchell, M. Willis, S.P. Armes, M.A. Khan, M.J. Percy and C. Perruchot, *Planet. Space Sci.*, 2002, **50**, 1025.

15. B.J. Goldsworthy, M.J. Burchell, M.J. Cole, S.F. Green, M.R. Leese, N. McBridge, J.A.M. McDonnell, M. Muller, E. Grun, R. Srama, S.P. Armes and M.A. Khan, *Adv. Space Res.*, 2002, **29**, 1139.
16. B.J. Goldsworthy, M.J. Burchell, M.J. Cole, S.P. Armes, M.A. Khan, S.F. Lascelles, S.F. Green, M. Müller, J.A.M. McDonnell, R. Srama and S.W. Bigger, *Astron. Astrophys.*, 2003, **409**, 1151.
17. (a) M. Gill, J. Mykytiuk, S.P. Armes, J.L. Edwards, T. Yeates, P.J. Moreland and C. Mollett, *J. Chem. Soc., Chem. Commun.*, 1992, 108; (b) M. Gill, S.P. Armes, D. Fairhurst, S. Emmett, T. Pigott and G. Idzorek, *Langmuir*, 1992 **8**, 2178; (c) N.J. Terrill, T. Crowley, M. Gill and S.P. Armes, *Langmuir*, 1993, **9**, 2093.
18. J. Stejskal, P. Kratochvil, S.P. Armes, S.F. Lascelles, A. Riede, M. Helmstadt, J. Prokes and I. Krivka, *Macromolecules*, 1996, **29**, 6814.
19. (a) S. Maeda and S.P. Armes, *J. Colloid Interface Sci.*, 1993, **159**, 257; (b) S. Maeda and S.P. Armes, *J. Mater. Chem.*, 1994, **4**, 935.
20. S. Maeda, M. Gill, S.P. Armes and I.W. Fletcher, *Langmuir*, 1995, **11**, 1899.
21. (a) C. Barthet, A.J. Hickey, D.B. Cairns and S.P. Armes, *Adv. Mater.*, 1999, **11**, 408; (b) M.J. Percy, C. Barthet, J.C. Lobb, M.A. Khan, S.F. Lascelles, M. Vamvakaki and S.P. Armes, *Langmuir*, 2000, **16**, 6913.
22. (a) J.I. Amalvy, M.J. Percy, S.P. Armes, C.A.P. Leite and F. Galembeck, *Langmuir*, 2005, **21**, 1175; (b) A. Schmid, S. Fujii, S.P. Armes, H. Minami, N. Saito, M. Okubo, C.A.P. Leite and F. Galembeck, *Chem. Mater.*, 2007, **19**, 2435.
23. S. Harley, D.W. Thompson and B. Vincent, *Colloids Surf.*, 1992, **62**, 163.
24. D.R. Skuse, T.F. Tadros and B. Vincent, *Colloids Surf.*, 1986, **17**, 343.
25. P.F. Luckham, B. Vincent and T.F. Tadros, *Colloids Surf.*, 1983, **6**, 119.
26. J. Kissel and F.R. Krueger, *Rapid Commun. Mass Spectrom.*, 2001, **15**, 1713.
27. (a) C. Barthet, S.P. Armes, S.F. Lascelles, S.Y. Luk and H.M.E. Stanley, *Langmuir*, 1998, **14**, 2032; (b) C. Barthet, S.P. Armes, M.M. Chehimi, C. Bilem and M. Omastova, *Langmuir*, 1998, **14**, 5032.
28. (a) M.A. Khan and S.P. Armes, *Langmuir*, 1999, **15**, 3469; (b) M.A. Khan, S.P. Armes, C. Perruchot, H. Ouamara, M.M. Chehimi, S.J. Greaves and J.F. Watts, *Langmuir*, 2000, **16**, 4171.
29. M.J. Burchell and S.P. Armes, unpublished results.
30. (a) R. Srama and E. Grun, *Adv. Space Res.* 1997, **20**, 1467; (b) see also http://www.jpl.nasa.gov/cassini/
31. (a) S.W. Kieffer, R. Lopes-Gautier, A. McEwen, W. Smythe, L. Keszthelyi and R. Carlson, *Science*, 2000, **288**, 1204; (b) J.R. Spencer, K.L. Jessup, M.A. McGrath, G.E. Ballester and R. Yelle, *Science*, 2000, **288**, 1208.
32. S. Fujii, S.P. Armes, S. Warren, S.L. McArthur, R. Devonshire, R. Jeans, M.J. Burchell, F. Postberg and R. Srama, *Chem. Mater.*, 2006, **18**, 2758.
33. (a) S. Kempf, R. Srama, F. Postberg, M. Burton, S. F. Green, S. Helfert, J.K. Hillier, N. McBride, J.A.M. McDonnell, G. Moragas-Klostermeyer, M. Roy and E. Grun, *Science*, 2005, **307**, 1274; (b) S. Kempf, R. Srama, M. Horanyi, M. Burton, S. Helfert, G. Moragas-Klostermeyer, M. Roy and E. Grun, *Nature*, 2005, **433**, 289.

Chapter 4

From Novel Monodisperse "Silicone Oil"/ Water Emulsions to Drug Delivery

Clive A. Prestidge

IAN WARK RESEARCH INSTITUTE, ARC SPECIAL RESEARCH CENTRE FOR PARTICLE AND MATERIAL INTERFACES, UNIVERSITY OF SOUTH AUSTRALIA, MAWSON LAKES, SA 5095, AUSTRALIA

Abstract

Brian Vincent and co-workers developed synthetic methods for the preparation of monodispersed "silicone oil" in water emulsions and further to crosslink these droplets into microgel particles. These highly versatile experimental systems have proven to be excellent novel model colloids which have inspired numerous subsequent studies over the last decade or more. In this chapter examples of colloid and interfacial investigations of monodisperse "silicone oil"/water emulsions are reported; a particular focus is on the ability to control droplet deformability and its impact on interfacial properties and interactions. The following areas are highlighted.

(i) Interaction forces on single silicone oil droplets determined using colloid probe atomic force microscopy (AFM) and the determination of droplet deformation on a nanometre scale and nanorheological properties of cross-linked droplets.
(ii) Rheological studies on concentrated emulsions of silicone droplets with different levels of cross-linking.
(iii) Adsorption of poly(ethylene oxide)–poly(propylene oxide)–poly(ethylene oxide) copolymers at the silicone droplet interface: the influence of polymer architecture and the droplet cross-linking level on the adsorbed polymer layer structure.

New Frontiers in Colloid Science: A Celebration of the Career of Brian Vincent
Edited by Simon Biggs, Terence Cosgrove and Peter Dowding
© The Royal Society of Chemistry 2008

(iv) Interaction studies of silica nanoparticles at the silicone droplet interface: the influence of nanoparticle wettability and the droplet cross-linking level on the adsorbed particle layer structure and droplet stability.

(v) Transport of poorly soluble molecules across the silicone droplet–water interface and the influence of nanoparticle layers. These transport studies have led to investigations of the loading and release of poorly soluble molecules (model drug compounds) from the silicone droplets and the observation that nanoparticle layers influence the release behaviour. These studies have formed the basis of a patented drug carrier technology that improves *in vitro* and *in vivo* drug delivery characteristics and is under development for applications for dermal and oral delivery.

4.1 Background

Brian Vincent has played a major role as a mentor in the author's career and provided direct inspiration for the silicone droplet studies presented in this chapter. After my chemistry degree at the University of Loughborough and an industrial training period working with Tharwat Tadros at the former ICI Plant Protection Division, I undertook a PhD at the University of Bristol under the joint supervision of Brian Vincent and Terry Cosgrove. My thesis studies on chemisorbing siloxane-based polymers at the solid–liquid interface were very much a marriage of the BV and TC areas of expertise and covered polymers in solution, polymers at interfaces and development of new experimental methods. Brian was also instrumental in facilitating my postdoctoral fellowship with John Ralston at the University of South Australia. I have been based in Adelaide since 1990 and my career has developed to the current position of research professor and sector coordinator for bio and polymer interfaces at the Ian Wark Research Institute; this is testament to the significance of Brian's influence on the careers of his research group members and many other young scientists.

4.2 Introduction

In 1994 Obey and Vincent[1] published a paper on novel monodisperse "silicone oil"/water emulsions where they reported methodology for the preparation of water-based emulsions of polydimethylsiloxane (PDMS) droplets through the ammonia-catalysed polymerisation of diethoxydimethylsilane (DEDMS):

$$n\text{Si}(\text{Me})_2(\text{OEt})_2 \xrightarrow{\text{OH}^-,\text{H}_2\text{O}} [-\text{Si}(\text{Me})_2-\text{O}-]_n + 2n\text{EtOH} \qquad (4.1)$$

This simple process enables monodispersed PDMS droplets to be prepared at volume fractions of a few percent and with diameters in the range 0.5–2 µm; droplet size is controlled by changing the ammonia to monomer ratio and the reaction time. In an extension of this work, Vincent and co-workers published work on inorganic "silicone oil" microgels[2] and showed that the inclusion of

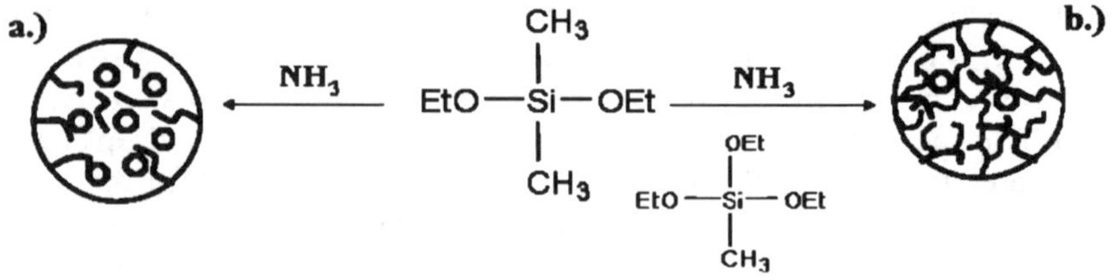

Figure 4.1   Schematic representation of silicone droplets (a) prepared from TEMS and non-cross-linked and (b) including DEDMS as a cross-linking monomer.

triethoxymethylsilane (TEMS), a trimeric monomer, resulted in internal cross-linking within the PDMS droplets. At TEMS : DEDMS ratios greater than 1 the droplets formed are substantially cross-linked and solid-like, but at lower levels of TEMS inclusion the droplets have viscoelastic properties, *i.e.* droplet deformability is controllable. The PDMS droplets and their composition are schematically represented in Figure 4.1.

^{1}H NMR and ^{29}Si NMR and gel permeation chromatography studies[1] have elucidated the composition of PDMS droplets prepared from DEDMS and have shown the presence of linear and cyclic PDMS oligomers with an average molecular weight of $420 \, \text{g mol}^{-1}$. The linear PDMS molecules are terminated with hydroxyl groups, which sit at the oil–water interface and undergo deprotonation as the pH is increased. In the simple case of liquid droplets where no cross-linker is present, the following protonation/deprotonation reaction scheme can be written:

$$-[O-Si(CH_3)_2]_n-OH_2^+ \underset{pK_{a1}}{\Leftrightarrow} -[O-Si(CH_3)_2]_n-OH + H^+$$
$$\underset{pK_{a2}}{\Leftrightarrow} -[O-Si(CH_3)_2]_n-O^- + 2H^+$$

(4.2)

These equilibria control the interfacial chemistry, the zeta potential behaviour and the colloid stability of the droplets. These droplets have isoelectric points in the pH range 2 to 4 and generally exhibit negative zeta potentials in aqueous solutions. A paper by Vincent and co-workers[3] reports further on the stability behaviour of the silicone oil emulsions and their film-forming properties.

4.3   Interaction Forces and Deformation of PDMS Droplets

The colloid probe atomic force microscopy (AFM) technique was originally developed to determine interaction forces between hard particles in solution,[4] and in recent years has been applied to bubbles,[5] soft polymer colloids,[6] droplets[7] and biological cells.[8] Gillies and co-workers[9–12] employed an AFM instrument in colloid probe mode to determine interaction forces between a silica sphere attached to the instrument cantilever and a surface-immobilised silicone droplet in aqueous solution. The approach and retraction of the colloid probe and a silicone droplet is schematically represented in Figure 4.2.

By varying the degree of internal cross-linking within the silicone droplets[2] their internal rheological properties (*i.e.* from Newtonian fluids to viscoelastic semi-solids) and hence deformability was systematically controlled and its influence on interaction forces determined.[11,12] For the range of equivalent droplets described Table 4.1, AFM interaction force data are shown in Figure 4.3.

It is noted that the data shown in Figure 4.3 are plotted against the arbitrary separation rather than actual separation. The reason here is that for AFM measurements on deformable systems there is no sharp transition between the contact and non-contact regions; this is because deformation occurs prior to contact due to the extended range of surface forces. Therefore, the zero of separation cannot be determined in the same way as for rigid bodies, *i.e.* from the region of constant compliance, where the colloid probe and solid particle move in a rigid coupled state.[4] That is, deformation of the silicone droplet results in additional variation of the surface separation not accounted for by the change in cantilever deflection minus the change in piezo position. For all

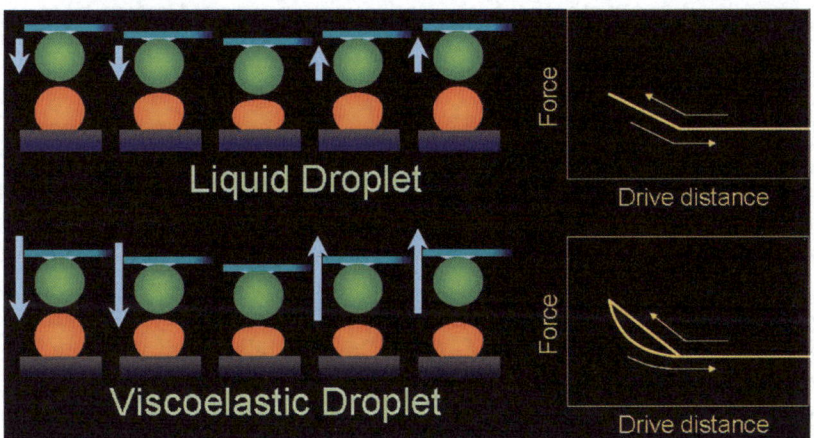

Figure 4.2 Schematic representation of a colloid probe AFM experiment to probe the interaction forces and deformation of silicone droplets. A high level of cross-linking introduces viscoelasticity and force curve hysteresis.

Table 4.1 Silicone droplets and their nanorheological parameters determined from AFM force data.

Droplet (% cross-linking)	A (50)	B (45)	C (40)	D (35)	E (30)
E_0 (MPa)	1.2	0.55	0.35	–	–
E_∞ (MPa)	0.8	0.4	0.3	–	–
τ (s)	0.12	0.07	0.05	–	–
k_{drop} (mN m^{-1})	250	140	83	59	47

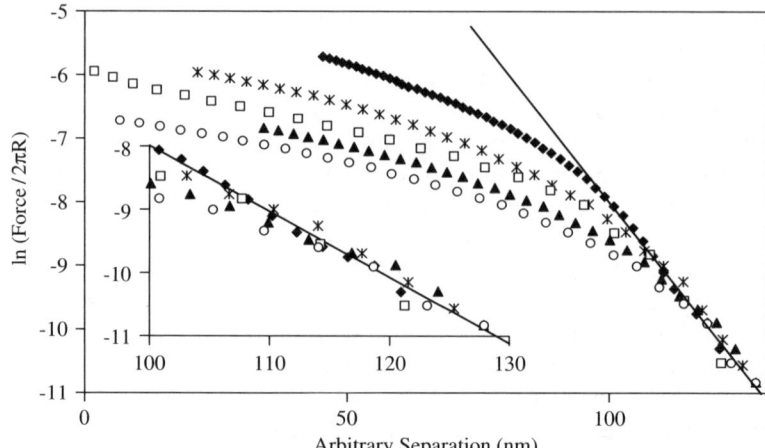

Figure 4.3   Approach force *versus* arbitrary separation curves for a glass probe
interacting with a droplet from emulsions A (◆), B (✳), C (□), D (▲), E
(○) in a background electrolyte solution of 1×10^{-3} M KNO$_3$ at
pH = 9.5. Solid line represents the Poisson–Boltzmann equation with a
Debye length of 9.6 nm. Inset shows an enlarged region where forces
decay at the same rate predicted by the Poisson–Boltzmann equation.
(Reproduced from ref. 12 with permission.)

levels of cross-linking forces are repulsive and increase monotonically. At
relatively large separations, interaction forces are well described by electrical
double layer interactions of rigid colloids, *i.e.* the Poisson–Boltzmann equation
(solid line in Figure 4.3), thus suggesting negligible deformation. On closer
approach, forces increase less rapidly with decreasing separation than expected
for hard-sphere behaviour and this is a result of deformation. Thus, it is postulated that droplets remain undeformed up to a critical load (F_{crit}), but deform
when the force exceeds this value. F_{crit} values correlate well with the level of
cross-linking and bulk material properties of the droplets.[11,12]

A method has been established to determine the nominal separation (defined
as the separation between surfaces as if they remain undeformed) based on the
observation that deformation is negligible for weak forces and the zero of
separation can be established by shifting the force data to coincide with the
rigid body interaction behaviour at large separations.[9] The rigid body interaction force (F) at large separation is well described by electrostatic repulsion
and can be predicted from the renormalised linear Poisson–Boltzmann equation for two spheres of an effective radius R_{eff} at a separation h:

$$F = 2\pi R_{eff} \varepsilon_r \varepsilon_0 \kappa^2 \left(\frac{4k_B T}{ze}\right) \gamma^2 e^{-\kappa h}; \quad \gamma = \tanh\frac{ze\psi}{4k_B T} \qquad (4.3)$$

where ε_r is the dielectric constant of water, ε_0 is the permittivity of free space,
κ^{-1} is the Debye length, k_B is Boltzmann's constant, T is the absolute

temperature, z is the valency of the electrolyte, e is the charge of an electron and ψ is the surface potential. Based on this approach, the interaction force data of Figure 4.3 are plotted against nominal separation in Figure 4.4; the retraction data are also included. Negative nominal separations correspond to inter-penetration of the undeformed bodies and give an indication of the level of deformation. Of further significance, force curve hysteresis is observed for the highly cross-linked silicone droplets (*i.e.* from emulsions A, B and C), but not for the liquid-like droplets (*i.e.* from emulsions D and E).

The extent of force curve hysteresis is dependent on the drive velocity and the extent of loading and is considered to be a result of a viscoelastic response, as is shown schematically in Figure 4.2. Similar viscoelastic responses have been reported in AFM studies of polymer materials[13] and biological cells,[8,14] but not for macroscopic droplets and bubbles. Upon deforming the highly cross-linked droplet, energy is stored elastically and dissipated viscously. Therefore, at relatively fast drive velocities, more energy is stored and droplets appear stiffer, and upon retraction more energy is dissipated and thus more hysteresis is observed. This behaviour is akin to a typical viscoelastic body, *e.g.* as described by the Maxwell model.

The force curves in Figure 4.4 were fitted by the relatively sophisticated viscoelastic theory developed by Attard[10,15,16] where separation, pressure and deformation are determined as a function of position and time, and a time-dependent elasticity parameter, $E(t)$. This is described by short-time, E_0, and long-time elasticity limits, E_∞, as well as a characteristic relaxation time, τ. The nanorheological parameters (E_0, E_∞ and τ) for the range of droplets under investigation are given in Table 4.1. For the highly cross-linked droplets from

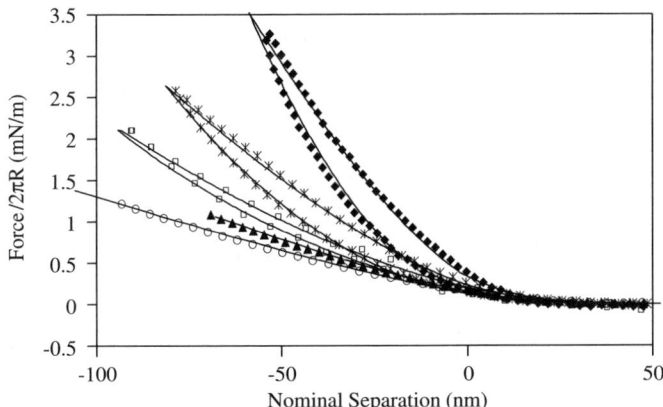

Figure 4.4 Force (approach and retraction) *versus* nominal separation for the inter-action of a glass probe and a droplet from emulsions A (◆), B (✳), C (□), D (▲), E (○) in a background electrolyte solution of 1×10^{-3} M KNO$_3$ at pH = 9.5. Solid black lines show theoretical force curve fits calculated using the nanorheological parameters in Table 4.1. (Reproduced from ref. 12 with permission.)

emulsion samples A, B and C the values for E_0 and E_∞ increase with increased cross-linking and are in good qualitative agreement with the bulk rheology of the PDMS within the droplets.[12] Droplets from emulsions D and E show no viscoelastic response and their deformation behaviour is adequately described by a single spring constant.

For liquid silicone droplets the influence of droplet size and surface tension (controlled by sodium dodecylsulfate (SDS) incorporation) on interaction forces and droplet deformation has been systematically studied.[11] The amount of deformation increases upon increasing the droplet radius, *i.e.* larger droplets are likely to be more deformable due to their reduced Laplace pressure. This is quantitatively confirmed by the observation that the spring constant is inversely proportional to the droplet radius (see Figure 4.5a). In addition, SDS inclusion has a pronounced effect on droplet deformability and the droplet spring constant is linearly related to the interfacial tension (see Figure 4.5b). A similar correlation has been reported for macroscopic decane droplets.[7]

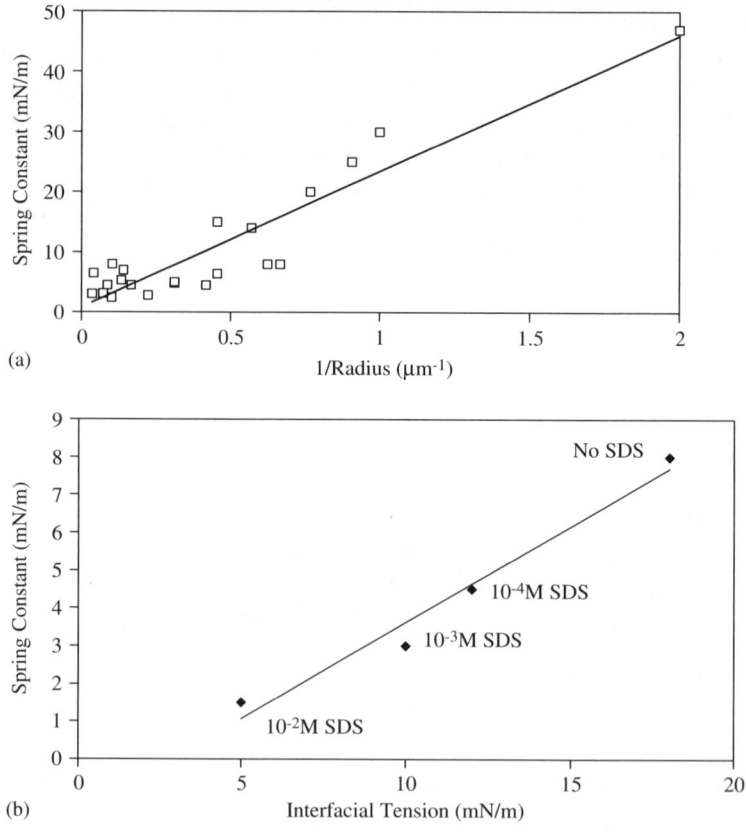

Figure 4.5 Spring constant for silicone droplets (with 30% cross-linked) plotted against (a) the inverse of droplet radius and (b) interfacial tension controlled by varying concentrations of SDS. (Reproduced from ref. 11 with permission.)

By systematically varying the amount of cross-linking within the silicone droplets, the influences of bulk rheology, size and interfacial tension on droplet interaction forces and deformability have been isolated. Highly cross-linked droplets are viscoelastic and the bulk rheology rather than interfacial effects controls their deformation. For liquid droplets, with low levels of cross-linking, the deformation is dominated by interfacial effects and the Laplace pressure controls droplet deformation. The interplay between the Laplace pressure and bulk rheology in controlling droplet deformation is well demonstrated using the silicone droplet system and gives insight into interdroplet interactions as well as interactions between bio-colloids or gas bubbles.

4.4 Rheological Behaviour of Concentrated Emulsions of PDMS Droplets

There are few reports of static and dynamic rheological studies on concentrated emulsions of monodispered droplets, particularly where the bulk rheological properties and hence deformability of the droplet are controlled. Saiki and co-workers[17,18] have used the silicone droplet system developed by the Vincent group[1,2] to explore the influence of droplet deformation on emulsion rheology. The initial challenge was to concentrate the silicone droplet system using centrifugation, retaining the droplet stability and monodispersity. Stable concentrated emulsions have been prepared over a wide volume fraction range ($0.1 \leq \phi \leq 0.72$) by the addition of SDS. Comparisons between silicone emulsions composed of "soft" (viscous) and "hard" (viscoelastic) droplets (*i.e.* with different cross-linking levels), and "hard sphere" silica suspensions have identified effects arising from the droplet deformability and dispersion structure on both the steady and dynamic rheology.

Rheological investigations of SDS-stabilised PDMS emulsions revealed that droplet deformability plays an important role in controlling the shear thinning behaviour. A "soft" emulsion was less shear thinning than a "hard" emulsion at low volume fractions, due to the high level of structural flexibility. At high volume fractions, however, the "soft" emulsion exhibited "extra" shear thinning behaviour, presumed due to lateral distortion of droplet structures.[18] This behaviour is represented schematically in Figure 4.6.

Relative viscosity (η_r) *versus* volume fraction (ϕ) curves for the silicone emulsions and comparisons with a colloidal silica suspension are reproduced in Figure 4.7. The Krieger and Dougherty[19] equation was developed to describe the viscosity behaviour of hard sphere suspensions:

$$\eta_r = \left(1 - \frac{\phi}{\phi_{max}}\right)^{-2.5\phi_{max}} \tag{4.4}$$

where ϕ_{max} is the maximum volume fraction (0.64 for random close packing and 0.74 for hexagonal close packing). It is clear from Figure 4.7 that silica

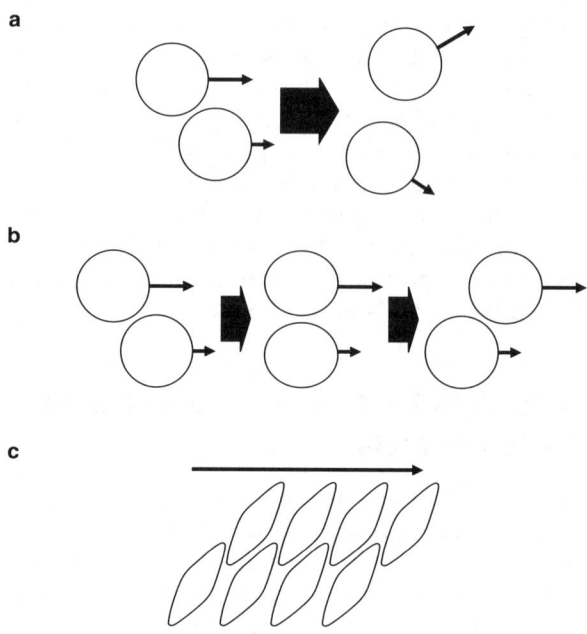

Figure 4.6 Schematic of the behaviour of particles or droplets under shear. (a) Hard
particles at moderate volume fractions; recoil of particles due to collision
causes drastic damage to the ordered structure created among particles,
resulting in a strong shear thinning behaviour. (b) Soft droplets at moderate volume fractions; due to deformation, the top droplet passes by the
bottom droplet without causing a significant change in the structure created among droplets, *i.e.* less shear thinning. (c) Soft droplets at high
volume fractions; droplets distort in an elongating manner, inducing extra
shear thinning behaviour. (Reproduced from ref. 18 with permission.)

suspensions follow the Krieger–Dougherty equation with a ϕ_{max} value of 0.68
over the entire volume fraction range. Silicone droplets behave akin to hard
spheres up to intermediate volume fractions, but deviate at specific volume
fractions due to deformation induced by some combination of hydrodynamic
forces and droplet collisions.

The relative viscosity of the emulsions of "hard" droplets begins to deviate
from the Krieger–Dougherty equation at $\phi = 0.58$, indicating the onset of
droplet deformation. Deformation allows droplets to pass by each other easily,
leading to an emulsion less resistant to the shear applied, therefore less viscous
than the hard sphere suspensions at the same volume fraction. The critical
deformation volume fraction of 0.58 coincides with the theoretically determined colloidal glass transition volume fraction, which originates from the
entrapment of droplets or particles in the "cage" formed by adjacent neigh-
bours. Shear will greatly disturb the "cage" structure due to enforced collisions,
resulting in high shear sensitivity. Emulsions of "soft" droplets deviate from the

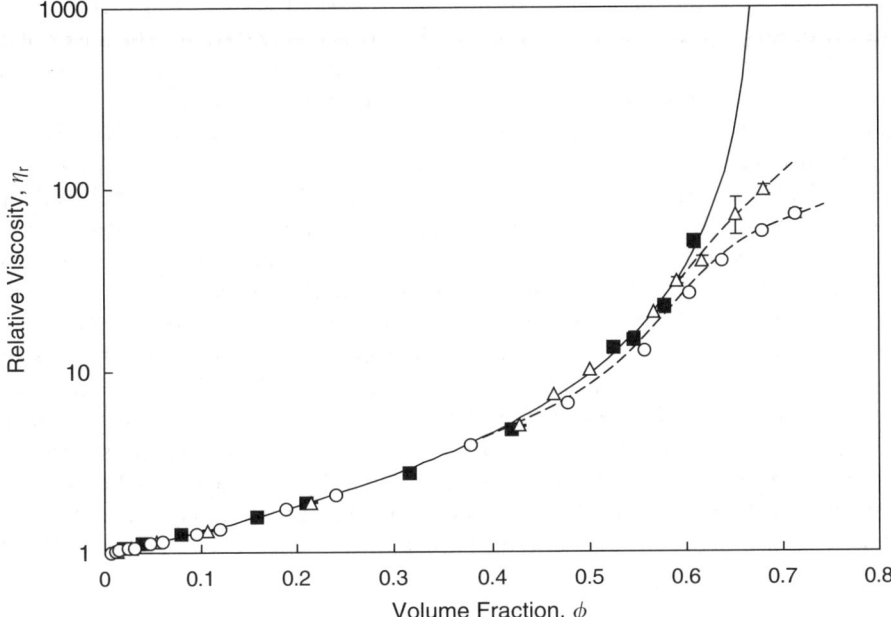

Figure 4.7   Relative viscosity (determined at a shear rate of $1000\,s^{-1}$) *versus* volume fraction profiles for: "soft" silicone droplets ($\bigcirc$), "hard" silicone droplets ($\triangle$) and silica particles ($\blacksquare$). Solid line represents the Krieger–Dougherty equation for $\phi_{max} = 0.68$. (Reproduced from ref. 18 with permission.)

Krieger–Dougherty equation at $\phi \approx 0.4$, well below the critical deformation volume fraction. Hydrodynamic interactions are considered to be responsible for deformation of the more liquid-like droplets.[18]

There is clearly a limitation in applying the Krieger–Dougherty equation to emulsions; the concept of critical deformation volume fractions observed for the deformability controllable silicone droplets may propel theory to predict the relative viscosity of emulsions over the entire volume fraction range.

4.5   Polymers at Silicone Droplet Interfaces

Non-ionic copolymers are extensively used as stabilisers for emulsions. A co-polymer's stabilising performance is controlled by its conformation at the oil–water interface, which is in turn influenced by its functionality and molecular architecture. Adsorbed copolymer conformations at solid–liquid interfaces have been extensively characterised.[20] However, equivalent studies on liquid–liquid interfaces or emulsion droplets are much less common. One reason for the lack of reported studies on polymers at droplet interfaces is that many of the conventional experimental approaches require measurements on colloids in the absence of a stabiliser, which is impractical for conventional emulsion droplets that readily coalesce without an adsorbed stabiliser. Furthermore, adsorption isotherm studies require specific surface area data for the adsorbent and these are not readily obtainable for emulsion systems; and conventional hydrodynamic thickness measurements (*e.g.* by light scattering) on adsorbed polymer layers are insensitive for micrometre-sized droplets. With these issues in mind, the stabiliser-free nature, controlled level of cross-linking and monodispersed nature of the silicone in water emulsion systems developed by the Vincent group[1,2] are well suited for studies to improve understanding of the relationship between copolymer characteristics, droplet penetrability and adsorbed copolymer conformation at the droplet–water interface.

Poly(ethylene oxide)–poly(propylene oxide)–poly(ethylene oxide) (PEO–PPO–PEO) triblock copolymers are highly effective emulsion stabilisers and have been extensively investigated. Studies by Barnes and co-workers[21,22] have been directed at PEO–PPO–PEO interaction behaviour with silicone droplets and have focused on the influence of copolymer structure and the role of interfacial penetration on the adsorption isotherms and adsorbed layer thickness. Adsorption isotherms for PEO–PPO–PEO copolymers at the liquid-like and cross-linked PDMS droplet–water interface are shown in Figure 4.8. To gain insight

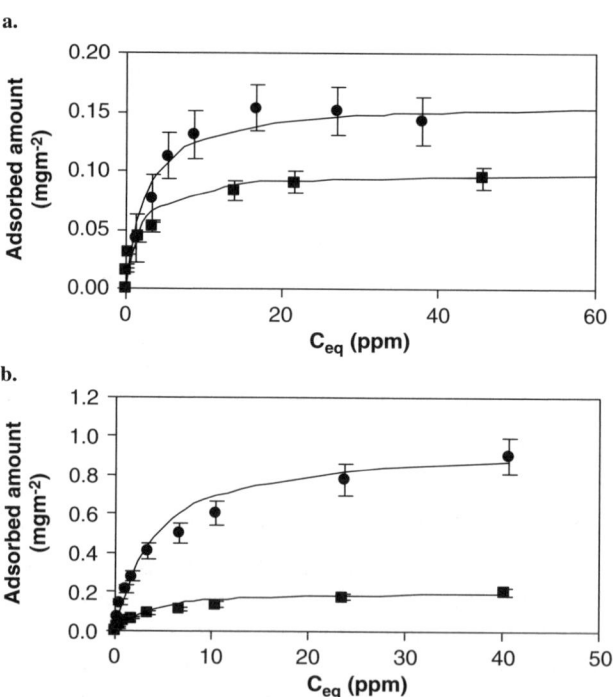

Figure 4.8 Adsorption isotherms for PEO–PPO–PEO block copolymers at the liquid-like (●) and cross-linked (■) PDMS droplet–water interfaces: (a) F108 and (b) P103. Lines are Langmuir model fits. (Reproduced from ref. 22 with permission.)

into the thermodynamics of adsorption at the droplet–water interface the adsorption isotherm data were fitted to the Langmuir model. Values for the free energy of adsorption (ΔG°_{ads}) were determined in the region of $-40\,kJ\,mol^{-1}$ and are in line with a physical adsorption mechanism. Plateau adsorbed amount (Γ_{max}) values are strongly dependent on the copolymer structure and are significantly lower for cross-linked droplets in comparison with liquid-like droplets, *i.e.* the equivalent surface area per adsorbed copolymer molecule is lower for liquid-like droplets (interestingly, this phenomenon is more pronounced for a copolymer with a longer PPO than PEO block length[21]).

The PEO–PPO–PEO adsorbed layer thickness (δ) was estimated from reduction of zeta potential due to movement of the plane of shear in the presence of the adsorbed non-ionic polymer.[21] The droplet zeta potential (ζ_1) is reduced to ζ_2 in the presence of an adsorbed polymer layer. By assuming that the interfacial charge, surface potential and charge distribution/movement of ions within the diffuse double layer are not significantly altered, the electrical double layer properties maybe described and enable δ to be determined:

$$\tanh\left(\frac{z_2 e \zeta_2}{4kT}\right) = \tanh\left(\frac{z_1 e \zeta_1}{4kT}\right) e^{-\kappa(\Delta-\delta)} \qquad (4.5)$$

where Δ is the distance from the surface to the shear plane (assumed to be equal to the thickness of the Stern plane, which is 0.4 nm) in the absence of polymer, κ is the inverse Debye length, z is the valence of the potential determining ions present, e is the electronic charge of the counter ions and kT is the thermal energy of the system. A similar approach has been used to determine the layer thickness of non-ionic homopolymers and copolymers adsorbed at solid particle–water interfaces with good success.

Examples of δ values for PEO–PPO–PEO copolymers at the surfaces of PDMS emulsion droplets are given in Figure 4.9. Errors in the values of δ from

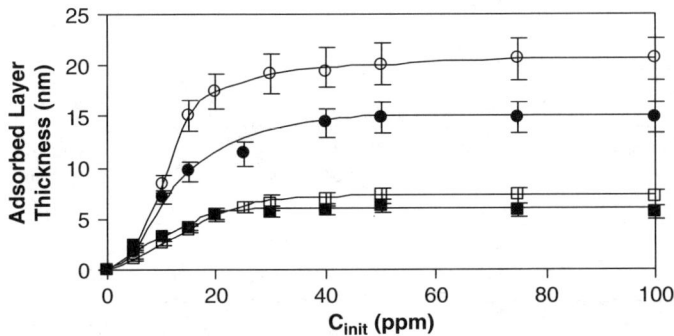

Figure 4.9 Influence of droplet cross-linking on the adsorbed layer thickness of PEO–PPO–PEO: F108, liquid (○) and cross-linked (●) droplets; P103, liquid (□) and cross-linked (■) droplets. (Reproduced from ref. 22 with permission.)

zeta potential variation (*i.e.* $< \pm 5\%$ for $\delta > 10$ nm, increasing to $\pm 20\%$ at lower δ values) are significantly less than those from other approaches. The determined δ values in the range 5 to 20 nm are in agreement with those determined from shifts in the viscosity *versus* volume fraction profiles of concentrated paraffin oil emulsions,[23] but greater than those generally determined using ellipsometry on flat solid surfaces. Ellipsometry is less sensitive to the presence of polymer tails and particle curvature may influence the layer thickness of an adsorbed copolymer.

For a wide range of copolymers a significant decrease in the layer thickness is apparent when the droplets are cross-linked.[21] The mobile nature and penetrability of the PDMS emulsion droplet–water interface are considered to be influential in controlling the adsorbed copolymer conformation and hence the adsorbed layer thickness. The neutron reflectivity studies of Phipps *et al.*[24] identified considerable penetration of the PPO block of a PEO–PPO–PEO (F127) across the oil–water interface (*e.g.* PEO segments extended 9 nm (~ 4.5 times the radius of gyration, R_g) into the water phase and the PPO segments 4 nm ($\sim 2.5 R_g$) into a hexane phase). Similar effects are envisaged for the silicone–water interface. Penetration of the PPO block through PDMS droplet–water interface results in the PPO blocks being shielded from unfavourable interactions with water, which reduces their intramolecular association with PEO and leaves the PEO blocks with more freedom to extend into solution. Furthermore, the surface area occupied by the PPO block at the interface decreases, resulting in increased steric repulsion between PEO blocks in the aqueous phase; the PEO blocks then take up a more extended conformation to minimise steric hindrance between neighbouring chains. These effects result in a more extended PEO block conformation for liquid-like silicone droplets in comparison with cross-linked droplets, with a greater adsorbed layer thickness and increased adsorbed amount; this is shown schematically in Figure 4.10.

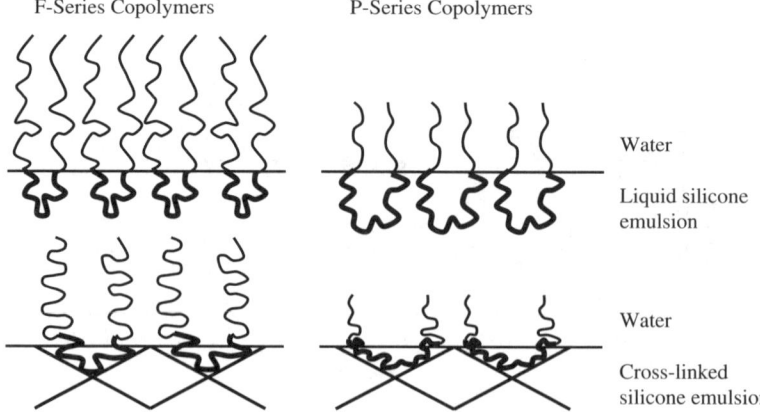

Figure 4.10   Schematic representation of the adsorbed PEO–PPO–PEO conformation at the PDMS liquid and cross-linked droplet–water interfaces. (Modified from ref. 21.)

These studies highlight the importance of interfacial penetration in controlling adsorbed copolymer conformations at emulsion droplet interfaces. Direct extrapolation of experimental findings obtained for solid particles to emulsion droplet interfaces may not therefore be fully justified. The silicone droplet system with different cross-linking levels is an effective model system for probing interfacial penetration of polymer segments and warrants further investigation.

4.6   Nanoparticles at PDMS Droplets

It is well established that emulsions are stabilised by solid particles, *i.e.* Pickering emulsions. Particle and nanoparticle layers at droplet interfaces have a number of advantageous properties over surfactant- or polymer-stabilised emulsions and the number of practical applications of Pickering emulsions and colloidosomes are on the increase. In a series of articles, Simovic and Prestidge[25–30] have employed the Vincent group's silicone droplets to probe the adsorption of hydrophilic and hydrophobically modified silica nanoparticles of ∼50 nm in diameter and to ascertain the role of nanoparticle layers in controlling droplet stability and interfacial transport processes. The stabiliser-free nature of the droplets enabled direct nanoparticle–"oil" interaction and the influence of droplet cross-linking enabled the role of interfacial penetration to be addressed for different nanoparticle types under different solution conditions.

Adsorption isotherms for hydrophilic silica nanoparticles at silicone droplets as a function of background electrolyte are presented in Figure 4.11. (Note that an equivalent hexagonally close-packed monolayer of hard spheres corresponds

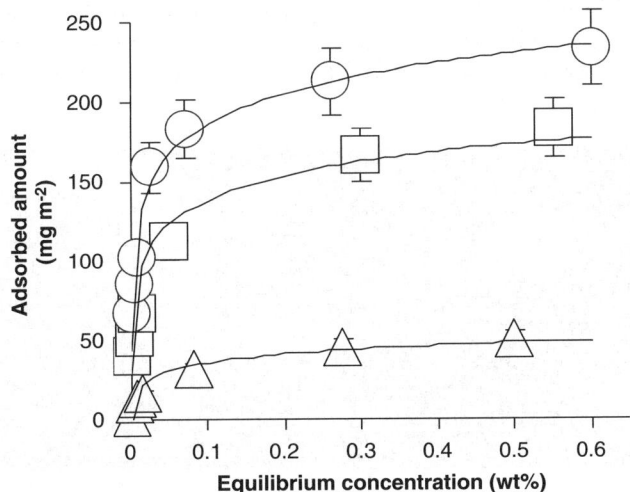

Figure 4.11   Adsorption isotherms for hydrophilic silica nanoparticles and liquid PDMS droplets as a function of salt concentration at pH = 9 and NaCl concentrations of 10^{-3} M (△), 10^{-2} M (□) and 10^{-1} M (○). (Reproduced from ref. 25 with permission.)

to $\sim 200 \, \text{mg m}^{-2}$ for 50 nm diameter silica particles.) The adsorption behaviour is "Langmuirian" and plateau surface coverage values correspond to their hard sphere radius + double layer thickness, *i.e.* lateral silica–silica interaction controls particle packing.[25] The free energy of adsorption (ΔG_{ads}) was determined to be approximately $-20 \, \text{kJ mol}^{-1}$ and indicative of a weak physical adsorption mechanism, *i.e.* dispersion forces and hydrogen bonding are sufficient to overcome the electrostatic repulsion between nanoparticles and droplets (a few kT). ΔG_{ads} and particle packing at the interface are only weakly influenced by pH, but significantly increased by salt addition. Droplet cross-linking reduced nanoparticle adsorption, but only at higher salt concentrations, where there is increased likelihood of silica particles wetting the PDMS and therefore interfacial penetration. Freeze-fracture scanning electron microscopy (SEM) images (Figure 4.12 shows examples) reveal that in the low to intermediate salt regime, individual silica nanoparticles are adsorbed at the droplet interface with negligible interfacial aggregation or penetration. Densely packed adsorbed particle layers are only observed when the double layer thickness is reduced to a few nanometres, *i.e.* at high salt concentrations. Hydrophilic nanoparticle layers enhance the colloid stability of emulsion droplets, but only weakly influence the coalescence kinetics.[27]

The adsorption isotherms for hydrophobically modified silica nanoparticles shown in Figure 4.13 are of higher affinity than those for hydrophilic nanoparticles. Furthermore, multilayer (non-Langmuirian) adsorption is observed[26] and a strong dependency on droplet cross-linking that is indicative of three-phase contact and interfacial penetration even at 10^{-4} M NaCl (see Figure 4.12). Densely packed adsorbed nanoparticle layers with interfacial aggregation are observed over a wide range of solution conditions. Both liquid and cross-linked PDMS droplets show strongly pH-dependent adsorption of hydrophobic nanoparticles in agreement with DLVO theory, but in contrast to hydrophilic silica nanoparticle adsorption. Of further note, interfacial saturation of

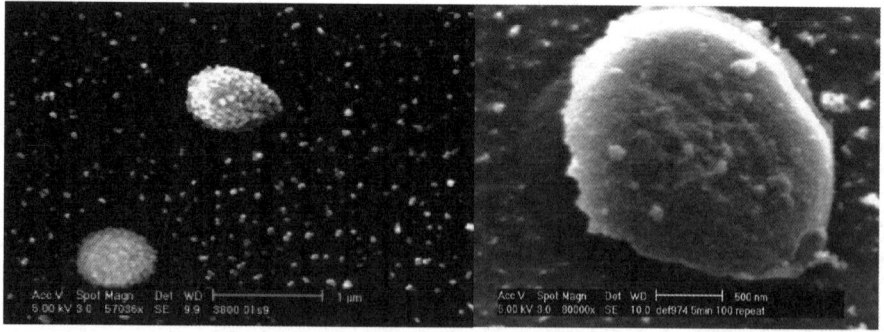

Figure 4.12 Freeze-fracture SEM images of liquid PDMS droplets with plateau coverages of silica nanoparticles. Left: hydrophilic silica nanoparticle at pH = 9 and 10^{-2} M NaCl. (Reproduced from ref. 25 with permission.) Right: hydrophobic silica nanoparticle at pH = 9 and 10^{-4} M NaCl. (Reproduced from ref. 26 with permission.)

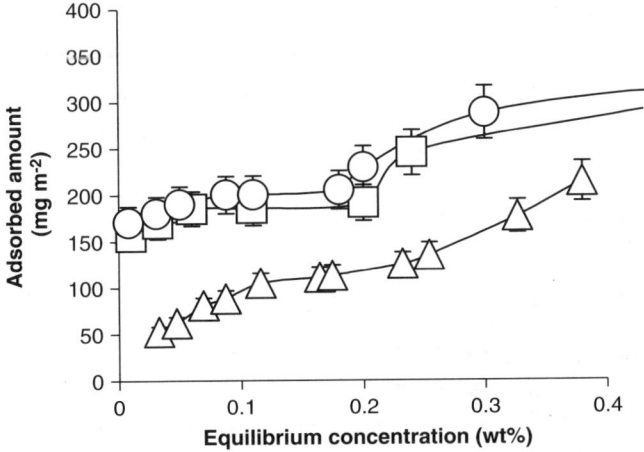

Figure 4.13   Adsorption isotherms for hydrophobic silica nanoparticles ($\theta \approx 117°$) and liquid PDMS droplets, at pH = 9 and NaCl concentrations of 10^{-4} M ($\triangle$), 10^{-3} M ($\square$) and 10^{-2} M ($\bigcirc$). (Reproduced from ref. 26 with permission.)

hydrophobically modified silica nanoparticles occurs at salt concentrations two orders of magnitude less than their critical coagulation concentration in solution.

Interfacially adsorbed hydrophobic nanoparticles reduce the colloid stability of silicone droplets, but significantly reduce droplet coalescence particularly at high surface coverages.[27] At intermediate coverage, partial coalescence of such systems results in the formation of a number of structures including tubes and dendritic structures, depending on the composition and hydrophobicity.

4.7 Towards a Drug Delivery System

Emulsions are highly effective carriers for lipophilic molecules and have many applications as drug delivery systems. However, drug molecules that are contained within oil droplets may partition readily to the oil–water interface and into the aqueous phase. These phenomena compromise effective encapsulation and emulsion stability. The Vincent group's model silicone emulsion system has been used to investigate the influences of droplet cross-linking and of adsorbed nanoparticle layers on the emulsion droplet to aqueous solution transport of lipophilic molecules.[29,30] Molecular transport can be significantly altered by either cross-linking or inclusion of nanoparticle layers; this behaviour is dependent on many factors, *e.g.* the level of molecule inclusion, the solution conditions and the type of nanoparticles.

The oil-to-water transport properties of the model lipophilic molecule dibutylphthalate (DBP) are strongly influenced by the presence of adsorbed layers of silica nanoparticles.[29,30] The transport kinetics are modified over several orders of magnitude as the nanoparticle layer microstructure (*i.e.* as controlled by the salt concentration), molecular loading within the droplets and droplet cross-linking level are varied. Figure 4.14 demonstrates that under

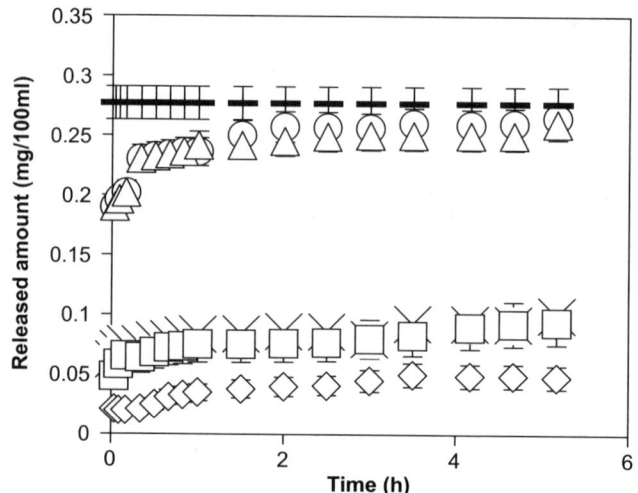

Figure 4.14   DBL release profiles from PDMS emulsions with low DBP loading under
sink conditions at 37 °C: uncoated emulsion (○) and systems containing
nanoparticles (△, □, ×, ◇). Dashed line represents 100% DBP dissolution (0.28 mg per 100 ml). (Reproduced from ref. 30 with permission.)

specific loading levels and for specific nanoparticle layer structures, significant
sustained release can be engineered. For highly structured layers of hydrophobic nanoparticles activation energies for release have been observed of
$\sim 600 \, \text{kJ mol}^{-1}$ and these are approximately ten times greater than for the
barriers introduced by non-ionic polymers.

In contrast, at higher DBP loading levels (total concentration greater than
the solubility level), both hydrophilic and hydrophobic nanoparticle layers
increase the rate and extent of dissolution compared with uncoated droplets
and pure DBP solutions (Figure 4.15). This effect is so pronounced that
supersaturated DBP solutions can be formed.

Nanoparticle coating of emulsion droplets is a useful strategy in solving
many drug formulation and delivery problems, *e.g.* long-term stability, protection of labile molecules and low orally dosed bioavailability of poorly soluble drugs. Ongoing *in vivo* studies have confirmed that nanoparticle-coated
emulsion carriers can improved the bioavailability of poorly soluble drugs for
both oral and dermal delivery.

4.8   Concluding Remarks

The novel monodisperse "silicone oil"/water emulsions that were pioneered by
Brian Vincent and his group have proven to be a highly valuable model colloidal system to investigate various colloid and interfacial phenomena. In
particular, the controllable cross-linking level and hence "silicone oil" droplets
of controllable deformability have enabled the influence of droplet

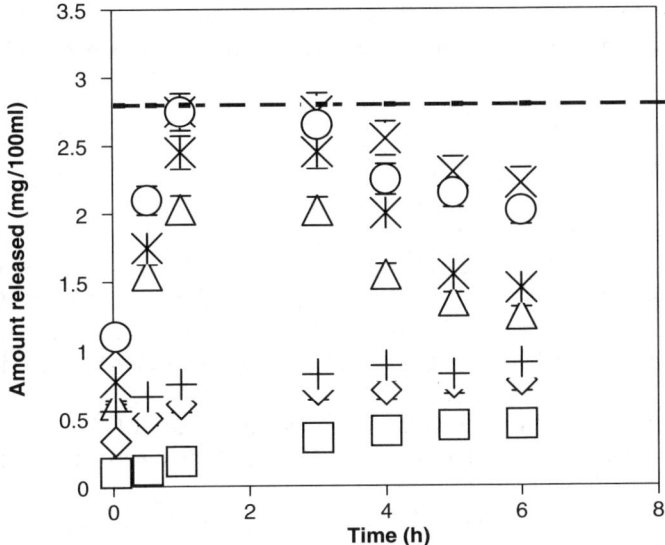

Figure 4.15 DBL release profiles from PDMS emulsions with high DBP loading under non-sink conditions at 37 °C: pure DBP (□), uncoated emulsion (◇) and systems containing nanoparticles (△, ○ ✳, ✕, +). Dashed line corresponds to 100% DBP dissolution/release (2.8 mg per 100 ml). (Reproduced from ref. 30 with permission.)

deformability and interfacial penetrability on colloidal interaction, polymer adsorption and nanoparticle adsorption to be investigated. This is not readily achievable with other colloidal systems.

Acknowledgements

Firstly Brian Vincent is sincerely acknowledged for inspiring this work and for having an enormous influence on the author's career. Dr Tim Barnes, Dr Graeme Gillies, Dr Yasushi Saiki and Dr Spomenka Simovic are acknowledged as the gifted and dedicated researchers who undertook the experimental studies on polymer interaction with droplets, colloid probe AFM of droplets, rheology of droplets and nanoparticle interaction with droplets, respectively—without their efforts this would not be possible. The Ian Wark Research Institute is acknowledged for long-term support and the Australian Research Council is acknowledged for funding.

References

1. T.M. Obey and B. Vincent, *J. Colloid Interface Sci.*, 1994, **163**, 454.
2. M.I. Goller, T.M. Obey, D.O.H. Teare, B. Vincent and M.R. Wegener, *Colloids Surf. A*, 1997, **123**, 183.

3. B. Neumann, B. Vincent, R. Krustev and H.-J. Müller, *Langmuir*, 2004, **20**, 4336.
4. W.A. Ducker, T.J. Senden and R.M. Pashley, *Nature*, 1991, **353**, 239.
5. M.L. Fielden, R.A. Hayes and J. Ralston, *Langmuir*, 1996, **12**, 3721.
6. P. Mulvaney, J.M. Perera, S. Biggs, F. Griesser and G.W. Stevens, *J. Colloid Interface Sci.*, 1996, **183**, 614.
7. S.A. Nespolo, D.Y.C. Chan, F. Grieser, P.G. Hartley and G.W. Stevens, *Langmuir*, 2003, **19**, 2124.
8. K. Bremmell, A. Evans and C.A. Prestidge, *Colloids Surf. B*, 2006, **50**, 43.
9. G. Gillies, C.A. Prestidge and P. Attard, *Langmuir*, 2001, **17**, 7955.
10. G. Gillies, C.A. Prestidge and P. Attard, *Langmuir*, 2002, **18**, 1674.
11. G.S. Gillies and C.A. Prestidge, *Adv. Colloid Interface Sci.*, 2004, **108–109**, 197.
12. G. Gillies and C.A. Prestidge, *Langmuir*, 2005, **21**, 12342.
13. C. Reynaud, F. Sommer, C. Quet, N. El Bounia and T.-M. Duc, *Surf. Interface Anal.*, 2000, **30**, 185.
14. E. A-Hassan, W.F. Heinz, M.D. Antonik, N.P. D'Costa, S. Nageswaran, C.-A. Schoenenberger and J.H. Hoh, *Biophys. J.*, 1998, **74**, 1564.
15. P. Attard, *Phys. Rev. E*, 2001, **63**, 061604.
16. P. Attard, *Langmuir*, 2001, **17**, 4322.
17. Y. Saiki and C.A. Prestidge, *Korean Australian J. Rheol.*, 2005, **17**, 191.
18. Y. Saiki, C.A. Prestidge and R.G. Horn, *Colloids Surf. A*, 2006, **299**, 65.
19. I.M. Krieger and T.J. Dougherty, *Trans. Soc. Rheol.*, 1959, **3**, 137.
20. G.J. Fleer, M.A. Cohen Stuart, J.M.H.M. Scheutjens, T. Cosgrove and B. Vincent, *Polymers at Interfaces*, Chapman & Hall, London, 1993 ch. 6.
21. T.J. Barnes and C.A. Prestidge, *Langmuir*, 2000, **16**, 4116.
22. C.A. Prestidge, T.J. Barnes and S. Simovic, *Adv. Colloid Interface Sci.*, 2004, **108–109**, 105.
23. Th.F. Tadros, *Colloids Surf. A*, 1994, **91**, 55.
24. J. Phipps, R. Richardson, T. Cosgrove and A. Eaglesham, *Langmuir*, 1993, **9**, 3530.
25. S. Simovic and C.A. Prestidge, *Langmuir*, 2003, **19**, 3785.
26. S. Simovic and C.A. Prestidge, *Langmuir*, 2003, **19**, 8364.
27. S. Simovic and C.A. Prestidge, *Langmuir*, 2004, **20**, 8357.
28. S. Simovic and C.A. Prestidge, *Aust. J. Chem.*, 2005, **58**, 664.
29. C.A. Prestidge and S. Simovic, *Int. J. Pharm.*, 2006, **324**, 92.
30. S. Simovic and C.A. Prestidge, *Eur. J. Pharm. Biopharm.*, 2007, **66**, 39.

Chapter 5

Polymers and Surfactants at Interfaces: Colloidal Lego for Nanotechnology

Simon Biggs

INSTITUTE OF PARTICLE SCIENCE AND ENGINEERING, SCHOOL
OF PROCESS, ENVIRONMENTAL & MATERIALS ENGINEERING,
UNIVERSITY OF LEEDS, LEEDS LS2 9JT, UK

5.1 Introduction

In the oft-quoted seminal address of Richard Feynman in 1959 entitled 'There's Plenty of Room at the Bottom', the idea of building complex systems from atom-by-atom or molecule-by-molecule synthesis was first described. This is generally considered to mark the 'birth' of nanotechnology.[1] Despite this obviously important event, it was not until the 1990s that scientists, engineers and funding agencies really absorbed the potential of nanotechnology and popularised the term.[2] It is my contention, however, that colloid scientists have been investigating and exploiting nanotechnology over a much longer period. Although the important dates are themselves open to discussion, the systematic study of nanoscale particulates can certainly be traced at least as far back as Faraday and his famous gold sol experiments.[3,4] A key aspect of nanotechnology is the idea that large complex functional objects can be built up by careful construction of atoms or molecules—the so-called 'bottom-up' fabrication approach. Again, almost without knowing it, colloid scientists have been exploiting just such approaches over many decades through the formulation of surfactant systems where simple surfactants can aggregate together into much larger, more complex and more functional objects. The benefits of these bottom-up approaches are seen in a range of commercial products such as shampoos or cosmetic creams where the aggregates provide important physical and chemical functionality through such features as rheology control as well as aesthetic properties like opalescence. The links between colloid science and

New Frontiers in Colloid Science: A Celebration of the Career of Brian Vincent
Edited by Simon Biggs, Terence Cosgrove and Peter Dowding
© The Royal Society of Chemistry 2008

nanotechnology are indeed many and may lead to the conclusion that colloid science was indeed the original nanotechnology! That, of course, is open to debate . . .

It was my great good fortune as an impressionable undergraduate student at Bristol in the early 1980s to be assigned a young(ish) member of staff, Brian Vincent, as my second year physical chemistry tutor. Brian's ability as a tutor and teacher is outstanding and I am sure I was not the only young student who was inspired to develop their career in science as a result of his influence. Anyway, a couple of years later after graduating from Bristol, Brian offered me the chance to study for a PhD under his tutelage. The subject of choice was 'Block Copolymer Microemulsions'—so started my ongoing interest in block copolymers, the colloidal aggregates that they form and the possible uses and applications of these systems.[5]

In my PhD work, we synthesised the block copolymers using living anionic polymerisation methodologies.[6] Whilst this approach allows the production of extremely low polydispersity index polymers, frequently as low as 1.01, there are a number of limitations. These include a limited range of monomer types for which this reaction methodology is applicable which limits the range of co-polymer chemistries available. Also, the reactions are extremely sensitive to impurities and so conditions of extreme cleanliness and the complete absence of water are necessary to ensure the reaction proceeds as expected. Despite these difficulties, we were able to realise a range of block copolymers that were capable of acting as stabilisers for microemulsion-like dispersions; the advantages of using block copolymers being the relative tolerance of changes in temperature and background salt.[7,8] Whilst the systems we produced had a number of beneficial features, it was apparent that the synthetic method was a significant limitation to the manufacture of copolymers that might have subsequent 'real-world' applications. Nonetheless, this provided a significant spur to my career-long interest in copolymers and their application to colloid technology.

The use of copolymer aggregation as a method to control aqueous solution properties led me next to investigate associative thickeners and their possible use as rheology modifiers.[9–13] In this work, a micelle-mediated reaction process was used to allow the formation of hydrophobically modified polyacrylamides using an aqueous free radical synthesis procedure. Whilst the materials produced were indeed capable of imparting desired rheology effects to aqueous systems, it was also apparent that the use of standard free radical synthesis resulted in poorly controlled polymers both in terms of the molecular weight and the chemical structure polydispersity of the samples. Their potential as well-defined building blocks for complex higher order structures must therefore be seen as limited. Nonetheless, one potential application of these materials as thickeners for explosive emulsion formulations has been exploited by Orica Explosives; the acrylamide backbone offered an important feature as it was stable in the extreme salt environments ($>26\,M\,NH_4NO_3$) found in modern bulk explosives formulations.[14]

As a result of these early research projects, it was clear to me at that time that the opportunities to use block copolymers as building blocks for higher order functional materials were significantly compromised by the available synthesis protocols available at that time for copolymer molecules. This was especially true if one also considered the scale-up and manufacturing of such systems with the associated cost penalties for using a technique such as anionic polymerisation.

5.2 Self-assembly of Block Copolymers and the Formation of 'Smart Nanoparticles'

One benefit of having worked in Brian's group has been the large informal network of other 'escapees' from the Bristol laboratories who are now in academic and industrial research positions. One of these is Professor Steve Armes, who is now at Sheffield University. About 10 years ago, Steve and I agreed to collaborate in the area of water-soluble stimulus-responsive block copolymers. Steve's team had begun developing expertise in living radical polymerisation techniques and were synthesising a range of block copolymers that were capable of undergoing reversible micellisation in response to a simple stimulus such as a change in pH.[15–17] Typical data that illustrate this reversible aggregation are shown in Figure 5.1 along with a schematic illustration of the process.

Living radical polymerisation techniques such as atom transfer radical polymerisation (ATRP) and reversible addition fragmentation transfer polymerisation (RAFT) emerged in the mid-1990s and have proven to be very interesting alternatives to classic polymerisation approaches. These methods are now being

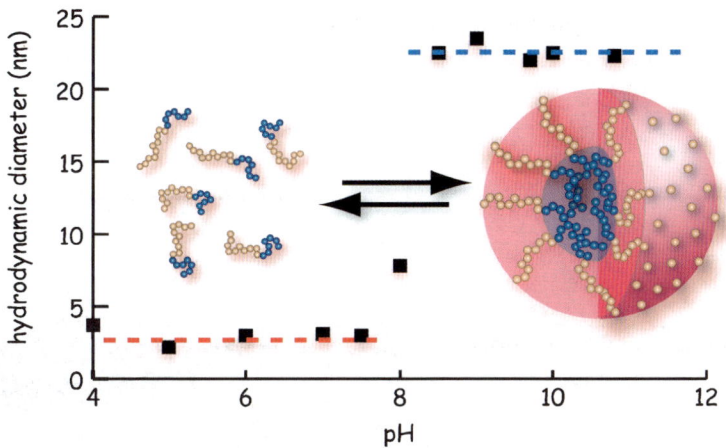

Figure 5.1   Typical data of the hydrodynamic diameter as a function of pH for a PDMA-PDEA copolymer in an aqueous solution of 0.01 M KNO$_3$. A schematic representation of the change from individual copolymer chains to a micellar aggregate at a critical pH (about 8 in this example) is also shown.

extensively employed since they combine the relative simplicity of free radical polymerisation approaches with many of the benefits of living anionic techniques such as control over the molecular weight and the polydispersity.[18–20] A direct result of these new synthetic procedures is the ability to design and synthesise a much wider range of polymers having controlled architectures and chemistry. It is possible therefore to consider the design of complex building blocks, such as functional copolymers, that can themselves aggregate into larger objects, such as block copolymer micelles. If these micelles can then self-organise into a higher level macroscopic structure, the functionality of the original copolymer can be translated into a higher order functional device or object; this is the essence of 'bottom-up' design and a key aim of nanotechnology! A schematic representation of these stages is shown in Figure 5.2.

In our work, we have been especially concerned with a class of stimulus-responsive AB block copolymers based on derivatives of poly(methyl methacrylate) chemistry. In particular, we have extensively investigated copolymers of the type poly[(2-dimethylamino)ethyl methacrylate]-*b*-poly[(2-diethylamino)ethyl methacrylate], which is often described by the acronym PDMA-PDEA. These copolymers are interesting primarily as a result of their weakly basic character which leads to a pH dependence of the solubility of both blocks. Previous investigations have shown that the degree of protonation for the tertiary amine groups as a function of pH is the same for both blocks.[21] However, the subtle increase in the hydrophobicity of the PDEA block means that it becomes insoluble at a lower pH than the PDMA block. The consequence of this, for AB block copolymers of these materials, is the reversible formation of micelle aggregates described in Figure 5.1. The exact pH where this aggregation occurs is somewhat dependent on the molecular weights of both blocks and the ratio of molecular weights within the copolymer. It is also worth noting here that the transition pH can be further tuned by altering

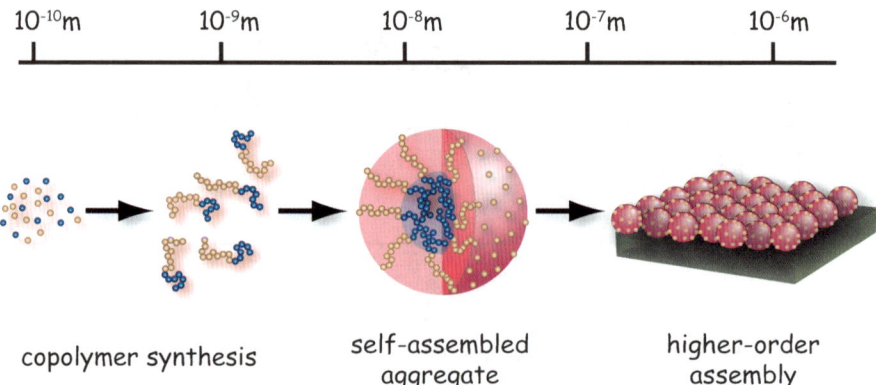

| 10^{-10}m | 10^{-9}m | 10^{-8}m | 10^{-7}m | 10^{-6}m |

copolymer synthesis   self-assembled   higher-order
                                aggregate   assembly

Figure 5.2   Schematic representation of a 'bottom-up' fabrication procedure involving functional block copolymers.

the substituent chemistries on the amine functionalities;[22] for example a PDMA-PDPA (where PDPA is poly[(diisopropylamino)ethyl methacrylate]) has a transition at a lower pH of around 7. The facile nature of this aggregate formation in response to a simple pH change as well as its reversibility suggests many possible applications for these copolymers in bulk solution such as pharmaceutical or agrochemical actives delivery vehicles or reversible stabilisers for emulsions and dispersions.

One significant benefit of these copolymers arises from the different reactivities of the tertiary amine residues. As a consequence of the differences in steric hinderance at these amine centres, it is possible to selectively quaternise only the PDMA blocks of the copolymers. Furthermore, this reaction is itself easily performed and stoichiometric, meaning that it is simple to introduce different degrees of permanent charge into the micelle aggregates.[17] One reason to add some charge, of course, is to completely eliminate any precipitation of the copolymer, even at high pH (>10) where the PDMA blocks would themselves become insoluble as a result discharging by deprotonation. The addition of permanent charge in this way can also be used to adjust the dimensions and properties of micelle aggregates. Examples of the change of properties for copolymers of this type as a result of the addition of charge are given in Table 5.1.

Whilst the bulk properties of these systems are undoubtedly of great interest, and such responsive micelle formation and break-up suggests many technology opportunities, a key interest of our work is how we might exploit these copolymer aggregates as building blocks for larger scale objects or processes where the functionality of the building block is important for the final use. One immediate idea was to examine whether these block copolymer micelles would adsorb to an interface and whether they might form organised surface structures. Associated with this, of course, are the effects of adsorption at an interface on their responsive properties.

Table 5.1   Solution properties of PDMA-PDEA copolymer[a] micelles at 25 °C.

Mean degree of quaternisation for PDMA chains (%)	Zeta potential (mV)[b]	D_H (nm)[b]	N_{agg}
0	+11	23.1	42 ± 3^c
10	+18	13.0	10 ± 1^c
			18 ± 1^d
50	+23	9.1	3.5 ± 1.5^c
			4.0 ± 0.5^d
100	+27	8.6	–

[a]Copolymer used here has a total molecular weight (M_w) of 19 100 g mol^{-1} and the PDMA-PDEA mole ratio is 79 : 21.
[b]In 10 mmol dm^{-3} KNO$_3$ aqueous solution at pH = 9. The error was estimated to be approximately 5% in the zeta potential measurement and 8% in the D_H measurement.
[c]In 10 mmol dm^{-3} Borax buffer at pH = 8.5.
[d]In 10 mmol dm^{-3} Borax buffer at pH = 9.5.

5.3   Adsorption of Block Copolymer Micelles at the Aqueous–Solid Interface

In a seminal paper in 1994, Manne and Gaub[23,24] demonstrated that simple surfactants such as sodium dodecylsulfate (SDS) or cetyltrimethylammonium bromide (CTAB) were capable of producing a rich variety of surface aggregate structures at the solid–liquid interface having similar dimensions to the aggregates seen in bulk solution, *i.e.* a few nanometres in diameter for a spherical micelle. This surprising (at the time) result suggested the possibility of producing self-assembled surface coatings with a high degree of order that could then be used as templates for further processing. Such an approach has been used by a number of authors to produce complex materials with similar order and across the same length scales to that of the original template.[25–27] Subsequent research has shown that a rich variety of surface structures may be formed such as full spheres, half spheres, tubules and bilayers.[28–30] The structure that is formed depends upon the type of surfactant employed as well as the features of the surface, such as charge density and roughness.[31,32] Whilst this work provides a useful fundamental benchmark and suggests a range of possible applications for self-ordered nanoscale surface coatings, one significant drawback is apparent and that is the requirement for bulk surfactant activity to ensure the presence of the structures. Put simply, rinsing the substrate results in immediate and complete breakdown of the structure. This would clearly limit the potential for using such systems in real-world applications or as template materials in multistage manufacturing processes.

One possible answer to this problem is to use polymers. It is well known that there is usually a significant hysteresis between the adsorption and desorption rates for polymers; once adsorbed at a substrate they are not easily rinsed off in many cases.[33–36] Hence, they can be kinetically trapped at the interface. A key question, of course, is whether the polymers can also produce the rich variety of surface aggregates seen with the small-molecule surfactants and whether these can be induced to self-assemble in the same way.

Over recent years, it has been demonstrated by many workers that deposition of block copolymers from selective organic solvents onto solid substrates can result in a rich variety of surface morphologies. Again, this work has been facilitated by the larger variety of copolymers available as a result of synthetic advances over the last decade. One example of the rich variety of surface aggregates available is shown in Figure 5.3.

Control over the type of structures produced is generally gained by adjusting copolymer features such as the overall molecular weight and the relative block lengths. External features such as the solvency of each block will also play a role. Such surface structures are clearly interesting as potential substrates for a wide range of applications. However, in most cases specialist deposition approaches such as spin coating are required; this, and the use of organic solvents, must be seen as non-ideal for large-scale manufacturing and/or complex multistage processing of these systems for potential applications.

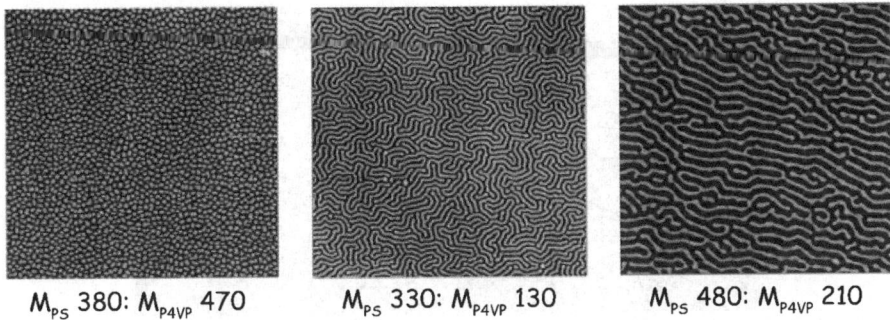

M_{PS} 380: M_{P4VP} 470 M_{PS} 330: M_{P4VP} 130 M_{PS} 480: M_{P4VP} 210

Figure 5.3   AFM images of ultrathin polystyrene-*b*-poly(4-vinylpyridine) (PS-*b*-P4VP) films deposited onto a flat substrate from a selective solvent. The three images show the different structures formed for different molecular weights of the two blocks in the copolymer, as detailed below each image. (Adapted from ref. 52.)

In our work, we have instead focused on aqueous copolymer systems that are capable of self-assembling at the solid–liquid interface to circumvent some of these issues. To date, facile control over a rich variety of structures, as seen above, has not been achieved from aqueous systems. Nonetheless, significant progress has been made and a number of systems which satisfy the self-assembly criterion from aqueous solution have been reported. In an earlier paper from our group, we demonstrated the formation of a close-packed array of spherical aggregates using the stimulus-responsive block copolymers described above.[37] Subsequent research has shown how the density of aggregates at the substrate depends on the balance of the micelle charge to the substrate charge.[38] For example, Figure 5.4 shows *in situ* soft-contact atomic force microscopy (AFM) imaging data for four different charge density copolymer samples on mica at a fixed pH. Under these conditions, where the surface charge is constant, the density of aggregates and the degree of close packing are clearly influenced by the charge on the copolymer micelles.

Analysis of the kinetics of adsorption using a combination of ellipsometry and quartz crystal microbalance measurements shows clearly that the final structures are strongly affected by the relative rates of copolymer adsorption and chain relaxation at the interface.[39,40] Not surprisingly, the higher the charge of the copolymer the faster its adsorption and the more rapid it collapses at the surface. The net result is that a single aggregate has a larger footprint and the density of aggregates is much reduced. Obviously, the film thickness is also affected by these changes in the density of aggregates and the degree of collapse. It has also been shown that the nature of the solid substrate in terms of its roughness and surface charge density also plays a role in the observed aggregate structures.[41] For example, a comparison of silica and mica is shown in Figure 5.5.

In related work, we have been exploring an alternative system for the facile preparation of surface coatings with nanoscale order from aqueous solution. This system consists of a polymerisable surfactant that is capable of forming

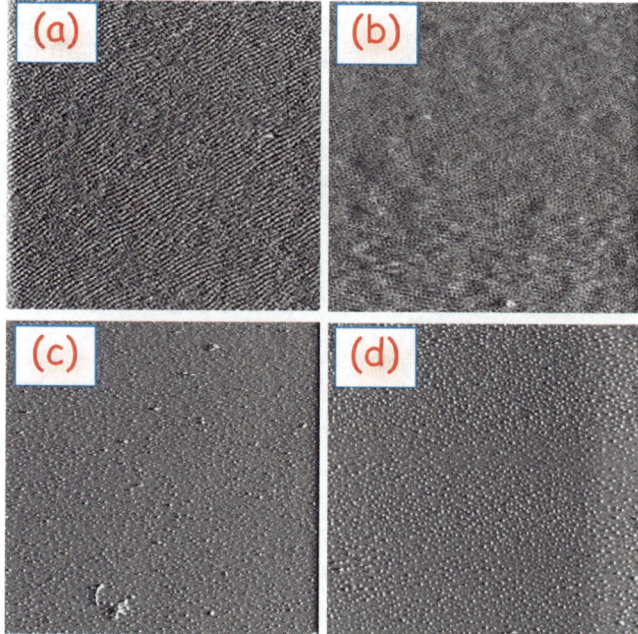

Figure 5.4   Soft-contact *in situ* AFM images of adsorbed PDMA-PDEA micelle layers as a function of the degree of permanent charge on the corona forming PDMA blocks, measured as the degree of quaternisation of the PDMA blocks: (a) 0%, (b) 10%, (c) 50% and (d) 100% quaternisation. Note that all images were collected after adsorption onto a pristine piece of mica from a 500 ppm copolymer solution in 0.01 M KNO_3 at pH = 9.

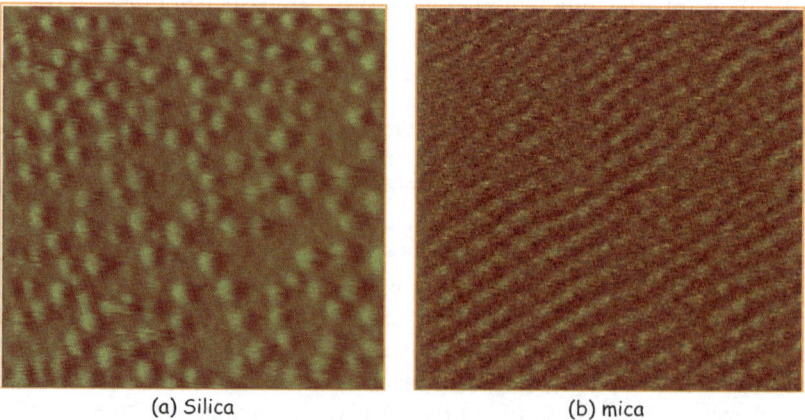

Figure 5.5   Soft-contact *in situ* AFM images of an adsorbed PDMA-PDEA micelle layer on (a) silica and (b) mica. These images were collected after adsorption of the micelles onto the clean substrates from a 500 ppm 0.01 M KNO_3 aqueous solution of the 0% quaternised copolymer.

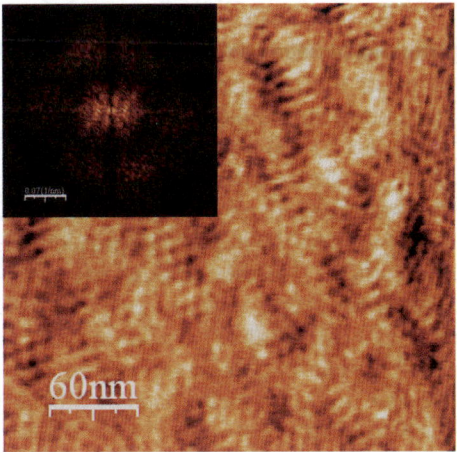

Figure 5.6   Soft-contact *in situ* AFM image of a pC18TVB layer adsorbed onto silica from a 0.03 mg ml⁻¹ aqueous solution. Inset shows the fast Fourier transform data for this image.

worm-like extended micelles in bulk solution. The system of choice, poly (stearyltrimethylammonium 4-vinylbenzoate) (p-C18TVB), retains the worm-like morphology after polymerisation.[42,43] Using similar approaches to those described above, we have shown that these polymers can form close-packed surface films spontaneously at the solid–liquid interface and the worm-like character is preserved (Figure 5.6).

Whilst not stimulus-responsive, these systems appear robust to washing and drying whilst retaining many of the characteristics of their small-molecule analogues.

5.4   Responsive and Functional Surfaces

As discussed above, we have over recent years investigated a range of stimulus-responsive block copolymers that can reversibly form micelles as a function of solution pH (see Figure 5.1). We have clearly shown that these copolymers will adsorb easily onto a substrate such as silica or mica from aqueous solution. Under the correct solution conditions, the result of this adsorption is the formation of a close-packed array of surface micelle aggregates at the solid–liquid interface (see Figures 5.4 and 5.5). Given that the copolymers of choice here are stimulus-responsive and it has been shown that their solution micelle aggregates can incorporate this responsive character, it was of interest to us to see how the higher level structures formed on the surface responded to the same stimulus.

Not surprisingly, exposure of the micelle aggregate coated surfaces to low pH was seen to result in a number of effects depending upon the exact nature of the copolymer chosen and the substrate used. In one case, for example, exposure of the micelle-coated surface to an acidic solution was seen to result in a slow irreversible swelling of the copolymer layer (Figure 5.7) giving a surface coating that was very reminiscent of fish scales.[37]

In other examples, the acidic pH was seen to cause a complete loss of the surface morphology.[44,45] These changes are monitored in real time using *in situ* AFM imaging techniques. Complementary force–distance data from AFM indicated, however, that the copolymer was still present at the interface despite there being no visible aggregates. Further analysis suggested that the copolymer was now in a brush-like form rather than the original micelle-like aggregates. Returning the solution pH to alkaline conditions was seen to reverse the morphology change and micelle-like aggregates were seen to re-form. Such a reversible open and close process for these micelles was likened to an anemone and is demonstrated schematically in Figure 5.8.[45]

Subsequently, we have shown that the reversibility of this process is also influenced by the substrate choice; mica-coated surfaces[45] readily showing reversibility whilst silica-coated ones do not.[44] Nonetheless, the opportunities suggested by these reversible open–close cycles for the micelles immobilised at the interface provide a significant impetus to work out how to produce such coatings on a wider range of surfaces. Further interest arises from the ability to tune the trigger point by using different chemistries as described above.[22]

It is clear that surfaces capable of exhibiting these kinds of morphology changes may have a number of possible applications. For example, the transition from an open to closed structure will be associated with a change in the interfacial friction properties and hence such surfaces might provide a system for stimulus-responsive control of hydrodynamics in narrow channels. Another possibility for their use could be as sensor substrates where the morphology

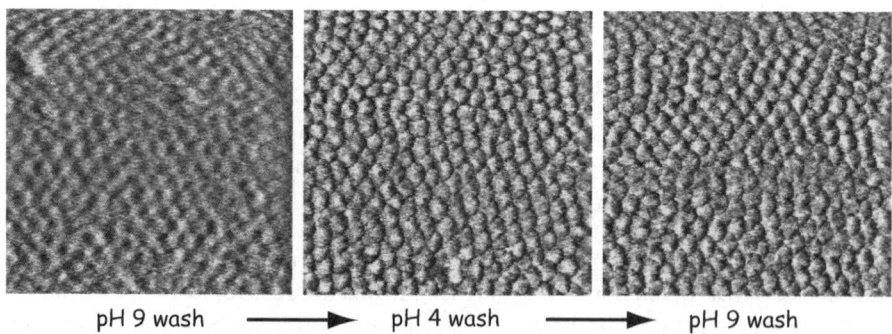

pH 9 wash ⟶ pH 4 wash ⟶ pH 9 wash

Figure 5.7 Soft-contact *in situ* AFM images of an adsorbed layer of 10% quaternised PDMA-PDEA micelles on mica adsorbed from a 500 ppm 0.01 M KNO$_3$ aqueous solution at pH = 9. Images were recorded after the fluid cell was flushed with (a) 0.01 M KNO$_3$ at pH = 9, (b) 0.01 M KNO$_3$ at pH = 4 and (c) 0.01 M KNO$_3$ at pH = 9 again.

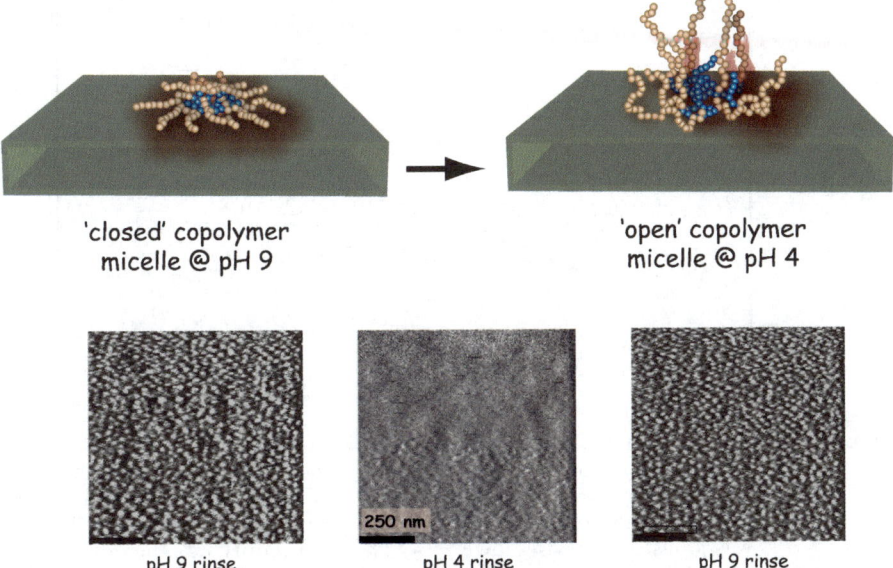

'closed' copolymer
micelle @ pH 9

'open' copolymer
micelle @ pH 4

pH 9 rinse   pH 4 rinse   pH 9 rinse

Figure 5.8   Schematic representation of the pH-responsive behaviour of an adsorbed PDMA-PDEA micelle at the oxide–water interface as the pH is adjusted from pH = 9 to 4. Corresponding *in situ* AFM images are also shown.

change is associated with some other detectable signal by selectively labelling the core blocks of the micelles, for example, with a chemical detector that responds to its environment. Another application area might be through the variable wetting properties that such coatings produce under different pH conditions as a result of the changing morphology and charge of the surface. An example of the contact angle, at the air–water–surface three-phase line, for a captive bubble is shown in Figure 5.9.[44]

Clearly, the surface morphology and charge variations across the pH range lead to dramatic variations in the wetting of the surface; these changes are also fully reversible over many cycles.

One important aspect of these reversible surfaces is their durability to repeated pH cycles, especially if they are to be used in environments where multiple reuse may be desired. A combination of ellipsometry techniques (optical reflectometry, OR) and quartz crystal microbalance (QCM) has proven invaluable in developing an understanding of the structure changes and durability of these layers across many pH cycles. Typical data are shown in Figure 5.10.

Initially, these data show the effects of rinsing the surface at both pH = 9 and 4. Clearly, during these rinse cycles any loosely bound copolymers are removed. However, it can be seen from the OR data that eventually a stable domain is reached where further desorption does not occur. Interestingly, in the region where the OR mass is stable as the pH is cycled the QCM recorded mass shows

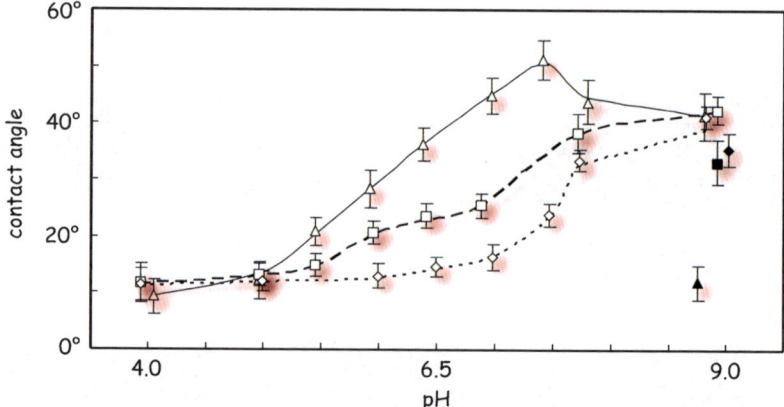

Figure 5.9   Variation in the contact angle of an adsorbed layer of PDMA-PDEA
micelles as a function of pH on silica. The initial copolymer films were
prepared from a 500 ppm copolymer solution at pH = 9. Contact angle
measurements were then made in the electrolyte rinse solution. The pH
was adjusted stepwise down to 4. The solution pH was then directly returned to 9 and the contact angle values were all found to be reversible
within error. The lines shown are guides to the eye, and do not represent a
mathematical fit to the data. Data are shown for three copolymer samples
at three levels of coronal quaternisation: 10% (triangles), 50% (squares)
and 100% (diamonds).

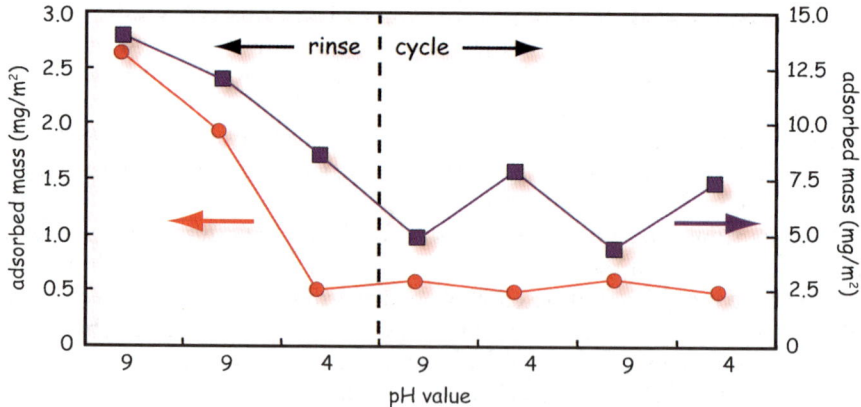

Figure 5.10   Change in the measured adsorbed mass as a function of pH recorded
using optical reflectometry (OR) and a quartz crystal microbalance
(QCM). The OR gives a measure of the true mass of copolymer adsorbed
whilst the QCM indicates the mass of polymer plus fluid contained
within the polymer layer. Comparison of the two data sets allows the
degree of hydration for the copolymer layer to be estimated.

large fluctuations. This is attributed to the structural changes in the adsorbed micelle layers described schematically in Figure 5.8. In the expanded state at pH = 4 the steric polymer layer can trap large amounts of water. Note that the QCM establishes a shearing motion and the plane of shear is likely to be in the outer parts of this steric layer; inside this plane of shear all material that moves with the crystal is coupled to it and is registered as an adsorbed mass. Hence, by combining the OR and QCM data, we can estimate the amount of liquid coupled in this layer under the different pH conditions, as well as inferring the changes in the adsorbed layer morphology. Importantly, these data when taken with the AFM images and force curve data indicate that the layers are robust to multiple pH cycles after some initial loss of loosely bound copolymer chains.

5.5 Multilayer Coatings, Particles and Capsules

Having fully characterised these copolymer films on simple flat substrates, we have recently turned our attention to more complex multilayer films on both flat substrates and particulates. The use of polymers to produce complex surface coatings by a simple layer-by-layer coating approach was first demonstrated by Decher and co-workers almost 20 years ago.[46] In this approach, a surface is coated by sequential immersion and rinsing in oppositely charged polyelectrolytes starting with one that will coat the surface of interest. Since this initial work, there have been many developments of this approach by using different polymer types, replacing one or more polymer layers with colloidal particles and coating particulates.[47–49] The attractiveness of this method for producing surface coatings comes from its flexibility in terms of the number of different species that may be used, the fact that it is water based and its ease of application. Inspired by this work, we have recently been exploring the development of multilayer films composed entirely of block copolymer micelles. A schematic illustration of the types of surface coating we hope to produce is shown in Figure 5.11.

In our work, we have continued to use PDMA-PDEA copolymers since they are extremely well characterised and they provide the cationic micelle building block. As well as this copolymer, we have also utilised an analogous zwitterionic copolymer, poly[(2-dimethylamino)ethyl methacrylate]-*b*-poly(methacrylic acid) (PDMA-PMAA). This copolymer forms cationic micelles at low pH and anionic ones at high pH.[17] Since the PDMA-PDEA copolymer forms micelles at any pH > 8, we are able to maintain stable micelle morphologies at pH = 9 allowing us to explore the production of multilayers under these conditions. In the simplest example, multilayers consisting of 5–6 layers of micelles have been deposited onto silica or mica flats. The adsorption of each layer has been monitored from the increase in mass as each subsequent layer is added. Interestingly, we have also been able to show that micelle structure can persist within the film as it is built up using *in situ* AFM imaging.[50]

The image shown in Figure 5.12 was collected by carefully scraping away one layer of micelles in a restricted area and then imaging the resultant surface over

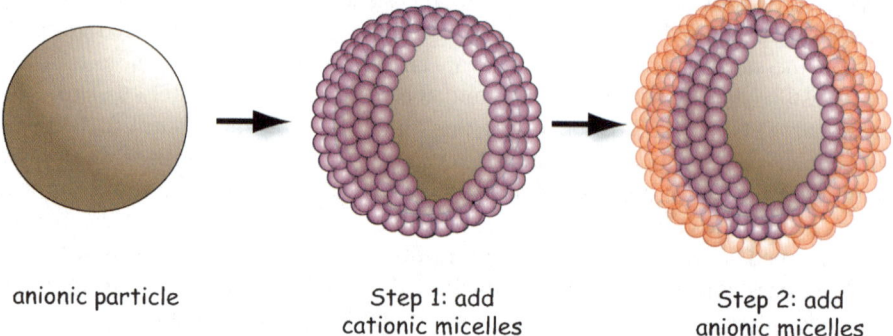

anionic particle   Step 1: add   Step 2: add
                                    cationic micelles   anionic micelles

Figure 5.11   Schematic representation of the idealised adsorption of two micelle layers
                  onto a spherical particle template showing the desired internal structure
                  of the layers.

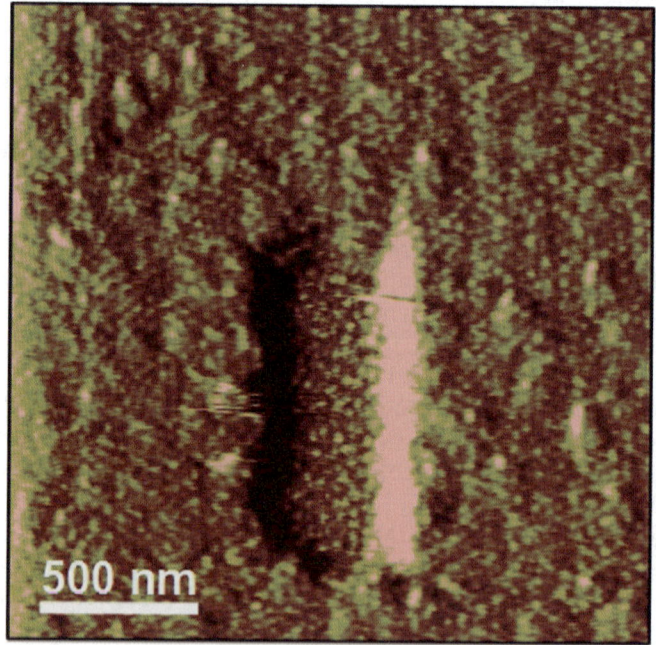

500 nm

Figure 5.12   *In situ* AFM image of a four-layer micelle-only film deposited onto a
                  silica substrate. The image clearly shows the structure recorded for the
                  outer layer and indicates the persistence of micelles in this layer. The
                  central hole was created by scanning this region with a large image force;
                  the hole is estimated to be one micelle layer deep and shows clear evi-
                  dence for the persistence of micelles in the lower layer (layer 3).

a wider area; clearly we can see micelle-like structures in two consecutive layers of the film. This important result suggests that we can produce a multilayer film exclusively of block copolymer micelles and the constituent micelles themselves retain their structures within the films. Given that these micelles are stimulus-responsive materials and that we effectively know where each micelle is within the film, we have produced a highly functional ordered three-dimensional film from the 'bottom up'!

Transfer of these coating ideas to particles has also been recently explored by us in related work.[51] Using particle templates is important for many reasons: they provide considerably more surface area than flat surfaces and so can increase availability of the films in applications; formulation of particle materials is generally easier for use in products, *etc.*; and the particles themselves may be present only as a template before subsequent dissolution resulting in the formation of a capsule. Our initial work in this area looks very promising. We have successfully demonstrated that micelle multilayer coatings can be easily prepared on small colloidal substrates and that these micelles are capable of carrying an active molecule within their cores (Figure 5.13).

We have now begun to explore whether the template particles can be dissolved away leaving a hollow capsule where the shell is composed entirely of block copolymers.

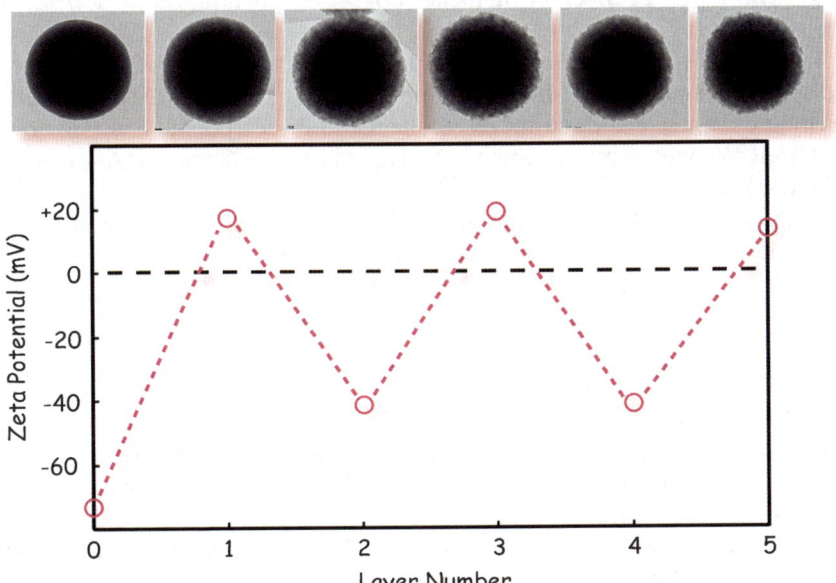

Figure 5.13 Data for the zeta potential as a function of micelle layer number for coated silica particles. Note that the zeta potential shows the expected charge reversal with each additional micelle layer as the cationic and anionic corona micelles are adsorbed sequentially. Corresponding transmission electron microscopy images of single particles coated with 0 through 5 layers are also shown.

5.6   Concluding Remarks

Molecular systems that are capable of self-assembling into larger scale objects are of central importance in the design and realisation of new functional materials using the bottom-up design principles of nanotechnology. Polymeric systems provide a number of key advantages over small molecules in this regard with respect to lifetimes of aggregates and other features. Recent advances in synthetic polymer chemistry have provided opportunities to design and synthesise a wide range of functional polymers and copolymers with a high degree of control over the polydispersity, both structure and molecular weight, of the samples produced. Using stimulus-responsive copolymer systems we have shown how complex three-dimensional surface coatings can be produced where the functionality of the building blocks is retained through to the final coating. A number of examples of this work are given here and demonstrate clearly the potential offered by copolymer surfactants as molecular building blocks for the design and manufacture of functional products.

Acknowledgements

I am extremely grateful to BNFL and the Royal Academy of Engineering for funding of my chair at Leeds in particle science and engineering. I am also grateful to the EPSRC for financial support of much of the work described here.

I would like to acknowledge the support and inspiration of a number of academic colleagues (Prof. Steven Armes and Dr. Erica Wanless) as well as current and former postdoctoral researchers and PhD students from my group (Dr. Grant Webber, Emelyn Smith, Dr. Kenichi Sakai, Timothy Addison and Pavlina Mantzana).

Finally, I acknowledge Brian Vincent for providing the inspiration to stick with science so many years ago. His support throughout my career has been remarkable, and without it I am sure I would not be where I am today!

References

1. R.P. Feynman, *J. Microelectromech. Syst.*, 1992, **1**, 60–66.
2. Foresight Nanotechnology Institute, http://www.foresight.org/.
3. R.J. Hunter, *Introduction to Modern Colloid Science*, Oxford University Press, Oxford, 1993.
4. D.J. Shaw, *Introduction to Colloid and Surface Chemistry*, Butterworth-Heinemann, 1992.
5. S. Biggs, Ph.D Thesis, University of Bristol, 1990.
6. S. Biggs and B. Vincent, *Colloid Polym. Sci.*, 1992, **270**, 505–510.
7. S. Biggs and B. Vincent, *Colloid Polym. Sci.*, 1992, **270**, 563–573.
8. S. Biggs and B. Vincent, *Colloid Polym. Sci.*, 1992, **270**, 511–517.
9. S. Biggs, A. Hill, J. Selb and F. Candau, *J. Phys. Chem.*, 1992, **96**, 1505–1511.

10. S. Biggs, J. Selb and F. Candau, *Langmuir*, 1992, **8**, 838–847.
11. S. Biggs, J. Selb and F. Candau, *Polymer*, 1993, **34**, 580–591.
12. F. Candau, S. Biggs, A. Hill and J. Selb, *Prog. Org. Coat.*, 1994, **24**, 11–19.
13. J. Selb, S. Biggs, D. Renoux and F. Candau, in *Hydrophilic Polymers*, ed. J.E. Glass, American Chemical Society, Washington, DC, 1996, *Adv. Chem. Ser.* Vol. 248, pp. 251–278.
14. S.T. Vittadello and S. Biggs, *Macromolecules*, 1998, **31**, 7691–7697.
15. F.L. Baines, S.P. Armes, N.C. Billingham and Z. Tuzar, *Macromolecules*, 1996, **29**, 8151–8159.
16. V. Butun, N.C. Billingham and S.P. Armes, *Chem. Commun.*, 1997, 671–672.
17. M. Vamvakaki, N.C. Billingham and S.P. Armes, *Polymer*, 1998, **39**, 2331–2337.
18. M. Kamigaito, T. Ando and M. Sawamoto, *Chem. Rev.*, 2001, **101**, 3689–3745.
19. K. Matyjaszewski and J.H. Xia, *Chem. Rev.*, 2001, **101**, 2921–2990.
20. S. Perrier and P. Takolpuckdee, *J. Polym. Sci. A: Polym. Chem.*, 2005, **43**, 5347–5393.
21. A.S. Lee, A.P. Gast, V. Butun and S.P. Armes, *Macromolecules*, 1999, **32**, 4302–4310.
22. V. Butun, S.P. Armes and N.C. Billingham, *Polymer*, 2001, **42**, 5993–6008.
23. S. Manne, J.P. Cleveland, H.E. Gaub, G.D. Stucky and P.K. Hansma, *Langmuir*, 1994, **10**, 4409–4413.
24. S. Manne and H.E. Gaub, *Science*, 1995, **270**, 1480–1482.
25. A.D.W. Carswell, E.A. O'Rear and B.P. Grady, *J. Am. Chem. Soc.*, 2003, **125**, 14793–14800.
26. H. Ding, C. Zhu, Z. Zhou, M. Wan and Y. Wei, *Macromol. Chem. Phys.*, 2006, **207**, 1159–1165.
27. J.P. Dong and G.Z. Mao, *Colloid Polym. Sci.*, 2005, **284**, 340–345.
28. W.A. Ducker and E.J. Wanless, *Langmuir*, 1999, **15**, 160–168.
29. H.N. Patrick and G.G. Warr, *Colloids Surf. A*, 2000, **162**, 149–157.
30. E.J. Wanless and W.A. Ducker, *Langmuir*, 1997, **13**, 1463–1474.
31. R. Atkin, V.S.J. Craig, E.J. Wanless and S. Biggs, *Adv. Colloid Interface Sci.*, 2003, **103**, 219–304.
32. G.G. Warr, *Curr. Opin. Colloid Interface Sci.*, 2000, **5**, 88–94.
33. J.C. Dijt, M.A.C. Stuart and G.J. Fleer, *Macromolecules*, 1992, **25**, 5416–5423.
34. J.C. Dijt, M.A.C. Stuart, J.E. Hofman and G.J. Fleer, *Colloids Surf.*, 1990, **51**, 141–158.
35. N.G. Hoogeveen, M.A.C. Stuart and G.J. Fleer, *J. Colloid Interface Sci.*, 1996, **182**, 133–145.
36. N.G. Hoogeveen, M.A.C. Stuart and G.J. Fleer, *J. Colloid Interface Sci.*, 1996, **182**, 146–157.
37. G.B. Webber, E.J. Wanless, V. Butun, S.P. Armes and S. Biggs, *Nano Letters*, 2002, **2**, 1307–1313.
38. G.B. Webber, E.J. Wanless, S.P. Armes and S. Biggs, *Faraday Discuss.*, 2005, **128**, 193–209.

39. K. Sakai, E.G. Smith, G.B. Webber, C. Schatz, E.J. Wanless, V. Butun, S.P. Armes and S. Biggs, *J. Phys. Chem. B*, 2006, **110**, 14744–14753.
40. K. Sakai, E.G. Smith, G.B. Webber, E.J. Wanless, V. Butun, S.P. Armes and S. Biggs, *J. Colloid Interface Sci.*, 2006, **303**, 372–379.
41. K. Sakai, E.G. Smith, G.B. Webber, C. Schatz, E.J. Wanless, V. Butun, S.P. Armes and S. Biggs, *Langmuir*, 2006, **22**, 5328–5333.
42. S. Biggs, S.R. Kline and L.M. Walker, *Langmuir*, 2004, **20**, 1085–1094.
43. S. Biggs, L.M. Walker and S.R. Kline, *Nano Letters*, 2002, **2**, 1409–1412.
44. K. Sakai, E.G. Smith, G.B. Webber, M. Baker, E.J. Wanless, V. Butun, S.P. Armes and S. Biggs, *Langmuir*, 2006, **22**, 8435–8442.
45. G.B. Webber, E.J. Wanless, S.P. Armes, Y.Q. Tang, Y.T. Li and S. Biggs, *Adv. Mater.*, 2004, **16**, 1794–1798.
46. G. Decher, *Science*, 1997, **277**, 1232–1237.
47. F. Caruso, R.A. Caruso and H. Mohwald, *Science*, 1998, **282**, 1111–1114.
48. F. Caruso, R.A. Caruso and H. Mohwald, *Chem. Mater.*, 1999, **11**, 3309–3314.
49. F. Caruso, H. Lichtenfeld, M. Giersig and H. Mohwald, *J. Am. Chem. Soc.*, 1998, **120**, 8523–8524.
50. E.G. Smith, G.B. Webber, K. Sakai, S. Biggs, S.P. Armes and E.J. Wanless, *J. Phys. Chem. B*, 2007, **111**, 5536–5541.
51. S. Biggs, K. Sakai, T. Addison, A. Schmid, S.P. Armes, M. Vamvakaki, V. Butun and G. Webber, *Adv. Mater.*, 2007, **19**, 247–250.
52. I.I. Potemkin, E.Y. Kramarenko, A.R. Khokhlov, R.G. Winkler, P. Reineker, P. Eibeck, J.P. Spatz and M. Moller, *Langmuir*, 1999, **15**, 7290–7298.

Chapter 6

Polymer Depletion: Recent Progress for Polymer/Colloid Phase Diagrams

Gerard Fleer

LABORATORY OF PHYSICAL CHEMISTRY AND COLLOID
SCIENCE, WAGENINGEN UNIVERSITY, 6703 HB WAGENINGEN,
THE NETHERLANDS

Abstract

A new theory for the phase behaviour of a mixture of colloids and non-adsorbing polymer is outlined. In such a mixture there is a depletion zone (thickness δ_s) around the colloids (radius a) which is void of polymer. Existing theory assumes δ_s to be equal to the radius R of the polymer coils. This is a fair approximation for the colloid limit (small $q = R/a$) but fails in the protein limit (large $q = R/a$) where δ_s becomes equal to the concentration-dependent correlation length ξ in semidilute solutions. The new theory includes a correct description for both limits and applies to intermediate size ratios ($q \approx 1$) as well.

6.1 Introduction

Brian Vincent has always had a keen interest in polymer depletion. This is the effect that nonadsorbing polymer is depleted from the region close to a surface: there is a depletion zone which is void of polymer. The width of this zone is called the depletion thickness. When two colloidal particles surrounded by depletion layers come close the depletion zones overlap, which induces an attraction between the particles caused by an unbalanced osmotic pressure: the outside osmotic pressure pushes the particles together because in the overlap region there is no polymer and no osmotic pressure. This attraction

New Frontiers in Colloid Science: A Celebration of the Career of Brian Vincent
Edited by Simon Biggs, Terence Cosgrove and Peter Dowding
© The Royal Society of Chemistry 2008

may lead to phase separation in a mixture of colloids and nonadsorbing polymer.

I estimate that about 50 papers by Brian and co-workers are in some way or another related to depletion. Many of them address the phase behaviour. It is interesting to note the changing terminology: when Brian started to work on this topic it was not yet clear whether the turbid drops formed in a polymer/colloid mixture were aggregated flocs or consisted of a true condensed (liquid or solid) phase. In an early paper[1] the aggregation process was denoted as *weak flocculation*, in later ones[2,3] it was called *reversible flocculation*, in one case[4] it is *controlled flocculation* and quite often[5-9] the term *depletion flocculation* is used. Gradually it emerged that depletion forces induce the formation of a real condensed phase,[10-12] and also Brian and co-workers started to use the term phase separation.[13-16] Regardless of the terminology, Brian's group was the first to come up with experimental data on the effects of depletion for well-defined systems.

In those days my own primary interest was in adsorbing polymers. However, when the Wageningen and Bristol groups joined forces in preparing the book *Polymers at Interfaces*[17] we had to include also depletion; the first author for this part was Brian. We had long discussions about 'pragmatic theories' *versus* '*ab initio* theories'.

Over the last decade or so Brian and myself more or less changed places. Brian's interests moved to, among other things, swelling and deswelling of microgels. I became more and more interested in analytical approximations for polymers at interfaces; these are easier for nonadsorbing polymers so I became involved more intensely in the field of depletion.

In this chapter I report on recent progress in the theoretical description of phase diagrams for mixtures of colloids and nonadsorbing polymer. Over the last few years Remco Tuinier (Jülich) and myself have been able to generalize the theory introduced by Lekkerkerker and co-workers.[18,19] This theory describes phase diagrams in terms of three parameters: the colloid volume fraction η and two ratios related to the polymer, *i.e.* $q = R/a$ and $y = \varphi/\varphi^*$. Here R is the radius of gyration of the polymer, a the radius of a colloid particle, φ the polymer volume fraction in the free volume (the volume not occupied by the colloids plus depletion layers) and φ^* the overlap concentration at which the polymer coils take up the entire solution volume. This overlap concentration separates the dilute regime (where the solution contains individual coils of radius R) from the semidilute regime (where the solution is a transient network of so-called blobs[20] of radius ξ). Hence, the polymer length scale varies from the high (chain length-dependent) value R in dilute solutions towards the much smaller (concentration-dependent) value ξ in the semidilute limit.

The original Lekkerkerker theory assumes that the depletion thickness δ_s around a colloidal sphere may be identified with R, and that the osmotic pressure Π is given by the van't Hoff law $\Pi = \varphi/N$, where N is the chain length. Both assumptions are reasonable for dilute polymer solutions. It turns out that for small q (colloidal particles much bigger than the coil volume) the polymer concentrations along the binodals are indeed below overlap ($y \ll 1$), so for this so-called *colloid limit* this model is a fair approximation.

However, for large q (coils bigger than the small protein-like colloids; this is the *protein limit*) the binodal polymer concentrations exceed the overlap concentration and the Lekkerkerker model breaks down. We generalize the model to include the correct semidilute limits $\delta_s \approx \xi$ and $\Pi \sim \xi^3$, where ξ is independent of chain length and depends on φ according to De Gennes' scaling law:[20] $\xi \sim \varphi^{-\gamma}$ (where in a good solvent $\gamma = 0.77$). Our generalized theory covers not only the two limits but also the transition region of intermediate q.

6.2 Theory

6.2.1 System and System Parameters

Figure 6.1 shows a schematic representation of the system considered in the Lekkerkerker theory: two phases with different colloid and polymer concentrations are in equilibrium with each other and with an external reservoir containing only the polymer solution. The colloid particles are taken as hard spheres with volume fraction η. The (external) polymer volume fraction φ fixes

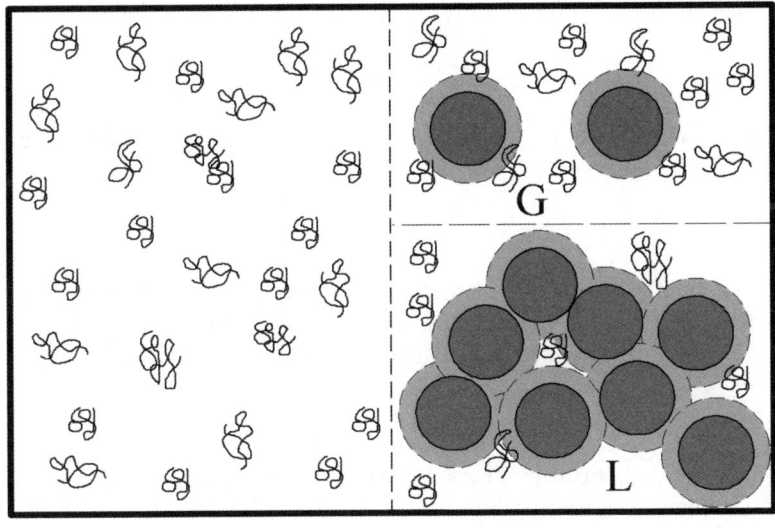

reservoir   system

Figure 6.1 Dilute colloidal gas G (colloid concentration η_G) in equilibrium with a colloidal liquid L (η_L) and with a reservoir containing only the polymer solution. The polymer volume fraction in this reservoir (and in the interstitial volume between colloids plus depletion layers) is denoted as φ. The depletion layers (thickness δ_s) are indicated as the grey halo around the particles (which have radius a).

the chemical potential of the polymer in the entire system. The example of Figure 6.1 depicts a GL equilibrium between a colloidal gas (G) and a colloidal liquid (L), but the condensed phase may also be a crystalline solid (S) so that a GS equilibrium occurs, and also LS coexistence may be found. It is well known[19,21,22] that for hard spheres without polymer ($\varphi = 0$) the only possible phase coexistence is LS, with coexistence concentrations $\eta_L^0 = 0.49$ and $\eta_S^0 = 0.54$.[22] In the presence of polymer the phase behaviour is richer and resembles that of simple one-component molecular substances: three types of two-phase coexistence (GL, GS and LS) occur and, as a consequence, also a GLS triple point with a three-phase equilibrium exists.

The phase behaviour is determined by three parameters. The first is, obviously, the colloid concentration η. It turns out that the ratio $\eta/(1 - \eta)$, which is the ratio between the volume occupied by the colloids and the volume not occupied by the colloids, occurs often in the equations. So we define a parameter f as

$$f = \frac{\eta}{1 - \eta}; \quad \eta = \frac{f}{1 + f} \tag{6.1}$$

The second parameter is the polymer volume fraction φ. Actually, only the ratio φ/φ^*, where φ^* is the overlap concentration, enters the equations. This overlap concentration depends on the chain length N and on the radius of gyration R of the polymer: $\varphi^* = N/v_{coil}$, with $v_{coil} = 4\pi R^3/3$; R is in units of the Kuhn length so R and v_{coil} are dimensionless. We thus define

$$y = \varphi/\varphi^* \tag{6.2}$$

The third parameter is the size ratio q between the coil radius R and the particle radius a:

$$q = R/a \tag{6.3}$$

Apart from the primary parameters η, y and q, we need some additional parameters which are a function of these three. The first is the size ratio q_s, which is a function of q and y. It is defined by

$$q_s = \delta_s/a \tag{6.4}$$

where δ_s is the depletion thickness around a colloidal sphere (*i.e.* the width of the grey depletion zones in Figure 6.1). In the colloid limit ($q \ll 1$) δ_s is not much different from R so that $q_s \approx q$; this is the situation described by the Lekkerkerker theory. In a slightly more refined picture curvature effects have to be taken into account because the depletion thickness around a sphere is somewhat smaller than that next to a flat plate. Then q_s is smaller than q, even in dilute polymer solutions.

A much more important effect is the compression of the depletion layers in concentrated polymer solutions: outside the colloid limit the polymer concentration is no longer in the dilute regime and δ_s is much smaller than R and becomes a function of the polymer concentration. Then also $q_s = q_s(q, y)$ is much smaller than q. We will see that in the protein limit $(q \gg 1)$ q_s becomes independent of q and is only a function of y. In this limit $\delta_s \approx \xi$, where $\xi \sim \varphi^{-\gamma}$ is the blob size (correlation length) in semidilute solutions.[20] Here $\gamma = 0.77$ is the De Gennes exponent in a good solvent, which is related to the Flory exponent $\nu = 0.588$ in $R \sim N^\nu$ through $(3\gamma - 1)(3\nu - 1) = 1$. It is easily shown[23] that the ratio $\xi/R = q_s/q$ scales as $y^{-\gamma}$. This implies $y \sim (q_s/q)^{-1/\gamma}$. Since ξ is independent of R in semidilute solutions, q_s is independent of q in the protein limit. It follows that the ratio $y = \varphi/\varphi^*$ is proportional to $q^{1/\gamma}$:

$$y \sim q^{1/\gamma} \quad \text{(protein limit)} \tag{6.5}$$

It is then convenient to define a new concentration parameter Y as

$$Y = yq^{1/\gamma} \tag{6.6}$$

In the protein limit Y becomes independent of q. Whereas in the Lekkerkerker theory the model was formulated in terms of q, η and y, we shall use the set q, η and Y because then the protein-limit results become more transparent.

A last parameter in the model is the free volume fraction α, which is the fraction of the (sub)system (G or L in Figure 6.1) available for the polymer: it is the fraction not occupied by the colloids plus depletion layers. Clearly, α in the L phase is much smaller than in the G phase (where it is close to 1). We use the standard scaled-particle result, which is[19,24]

$$\alpha = (1 - \eta) \exp(-Af - Bf^2 - Cf^3) \tag{6.7}$$

where the coefficients A, B and C depend only on q_s according to

$$A = (1 + q_s)^3 - 1; \quad B = 3q_s^2\left(q_s + \frac{3}{2}\right); \quad C = 3q_s^2 \tag{6.8}$$

6.2.2   Thermodynamics

The grand potential density ω of the colloid/polymer mixture is separated into a hard-sphere part ω^0 and a polymer contribution ω^p:

$$\omega = \omega^0 + \omega^p \tag{6.9}$$

For ω^0 we have to differentiate between the fluid (F) phases (either G or L) and the crystalline solid (S). Expressions are available in the literature:[25,26]

$$\omega^0 = \begin{cases} \eta[\ln \eta - 1 + 4f + f^2] & \text{F} \\ \eta[2.1306 - 3\ln(1/\eta - 1/\eta_{cp})] & \text{S} \end{cases} \tag{6.10}$$

where $\eta_{cp} = (\pi/6)\sqrt{2} = 0.741$ is the volume fraction at close-packing.

For the polymer contribution we modify the Lekkerkerker expression[19,24]

$$\omega^{\mathrm{P}} = - \int_0^{\Pi v} \alpha \, d\Pi v = - \int_0^Y \alpha \frac{\partial \Pi v}{\partial Y} \, dY \tag{6.11}$$

The parameter Πv, where $v = 4\pi a^3/3$ is the colloid volume, is the work to insert a bare colloidal particle into the polymer solution. In the colloid limit, where $\alpha = \alpha(q, \eta)$ does not depend on the polymer concentration, eqn (6.11) simplifies to $\omega^{\mathrm{P}} = -\alpha \Pi v$.

The remaining problem is to find expressions for $\alpha = \alpha(q_s, \eta)$, hence for $q_s(q, y)$, and for $\Pi v(q, y)$. Recently[27] we derived combination rules for the depletion thickness δ at a flat plate and for the osmotic pressure Π: $\delta^{-2} = \delta_0^{-2} + \delta_{sd}^{-2}$ and $\Pi = \Pi_0 + \Pi_{sd}$, where the indices 0 and sd refer to the dilute and semidilute limits, respectively. These limits are[28] $\delta_0 = 1.071R$ and[27,29] $\delta_{sd} = \xi = 0.50 \, \delta_0 y^{-\gamma}$ for δ; and $\Pi_0 = \varphi/N$ and $\Pi_{sd} = 1.62 \, \Pi_0 y^{3\gamma - 1}$ for Π.[27,30] Now q_s and $\partial \Pi v/\partial Y$ are simple functions of q and Y:

$$q_s = 0.866(q^{-2} + 3.95 Y^{2\gamma})^{-0.44} \tag{6.12}$$

$$\frac{\partial \Pi v}{\partial Y} = q^{-1/v} + 3.77 Y^{3\gamma - 1} \tag{6.13}$$

In eqn (6.13), $1/v$ may also be written as $3 - 1/\gamma$. The two limits of eqn (6.12) and (6.13) are immediately clear. In the colloid limit the Y terms may be omitted and $q_s = 0.866q^{0.88}$ is of order q; the difference between q_s and q is due to curvature effects. In this limit $\partial \Pi v/\partial Y = q^{-3+1/\gamma}$ or $\partial \Pi v/\partial Y = q^{-3}$, which is fully consistent with van't Hoff's law $\Pi = \varphi/N = y\varphi^*/N = y/v_{\mathrm{coil}}$ so that $\Pi v = yv/v_{\mathrm{coil}} = yq^{-3}$.

In the protein limit the q terms in eqn (6.12) and (6.13) are negligible: $q_s = 0.47 Y^{-0.68}$ and $\Pi v = 3.77 Y^{2.31}$ are independent of q and thus reach a constant level. In terms of $y = Yq^{1.3}$, with constant Y, this means that y scales as $q^{1.3}$, which agrees with eqn (6.5).

Equations (6.9)–(6.13) (in combination with eqn (6.7) and (6.8) for α) provide an analytical expression for the free energy of the system as a function of q, η and Y. Unlike in the Lekkerkerker theory this expression is valid for any q, and it includes the colloid and protein limits plus the crossover. The complete phase diagram now follows from standard thermodynamics. For example, binodal points for two-phase coexistence are computed from equal chemical potentials μ and pressures p in the two phases; these are given by

$$\mu = \frac{\partial \omega}{\partial \eta} = (1 + f)^2 \frac{\partial \omega}{\partial f}; \quad pv = -\omega + \eta \mu \tag{6.14}$$

Explicit analytical expressions for μ and pv (and for the first derivatives μ' and p' and second derivatives μ'' and p'' with respect to η) are available, but I do not give those here. Critical GL points are found from $\mu' = \mu'' = 0$ (or $p' = p'' = 0$); triple points follow from equal μ and p in three phases. Also the *critical endpoint* (see eqn (6.15)) is found immediately.

6.2.3 Some Illustrations

In this section I give two examples to illustrate some thermodynamic features. The first is a hard-sphere system without polymer; then μ^0 and $(pv)^0$ follow from eqn (6.14) by substituting $\omega = \omega^0$. Figure 6.2 gives μ^0 and $(pv)^0$ as a function of $f = \eta/(1 - \eta)$ for such a hard-sphere system, using eqn (6.10a) for a fluid (F) and eqn (6.10b) for a solid (S). It is well known that for hard spheres there is only FS demixing: computer simulations[22] give $\eta_F^0 = 0.494$ and $\eta_S^0 = 0.545$ for the coexistence compositions. The same result follows from $\mu_F^0 = \mu_S^0$ and $p_F^0 = p_S^0$, using eqn (6.10) and (6.14). The analytical coexistence condition is given by the rectangle in Figure 6.2: $\eta_F^0 = 0.492$, $\eta_S^0 = 0.542$ at chemical potential $\mu_0^0 = 15.463$ and pressure $(pv)_0^0 = 6.081$.

The second example is the dependence of μ and pv on the colloid concentration, now in the presence of polymer. I give a simplified picture for the colloid limit, omitting the Y terms in eqn (6.12) and (6.13). Then q_s and α do not depend on the polymer concentration, and $\mu = \mu^0 + \Pi vg$ and $pv = (pv)^0 + \Pi vh$. The functions $g = -\partial\alpha/\partial\eta$ and $h = \alpha - \eta\partial\alpha/\partial\eta$ are known (analytical) functions of q_s and η; the polymer concentration enters only through the term Πv in these expressions. Figure 6.3 gives $\mu(f)$ and $pv(f)$ for a fluid (G or L) phase and

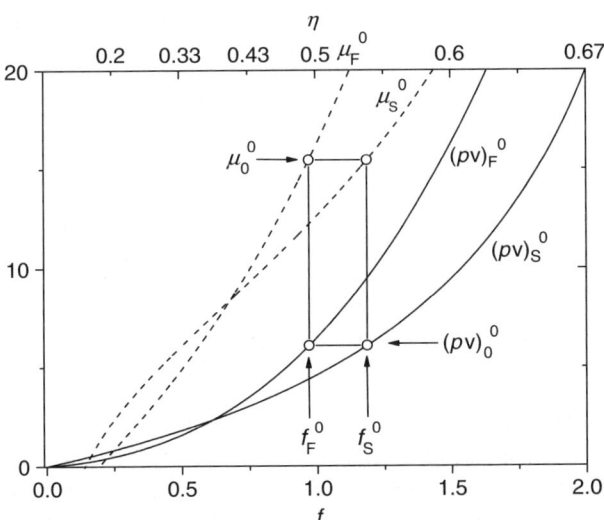

Figure 6.2 Hard-sphere chemical potentials μ_F^0 (fluid) and μ_S^0 (solid) and the corresponding $(pv)_F^0$ and $(pv)_S^0$ as a function of $f = \eta/(1 - \eta)$. The rectangle indicates the FS coexistence where $\mu = \mu_0^0 = 15.463$ and $pv = (pv)_0^0 = 6.081$, at concentrations $\eta_S^0 = 0.492$ ($f_F^0 = 0.970$) and $\eta_S^0 = 0.542$ ($f_F^0 = 1.185$).

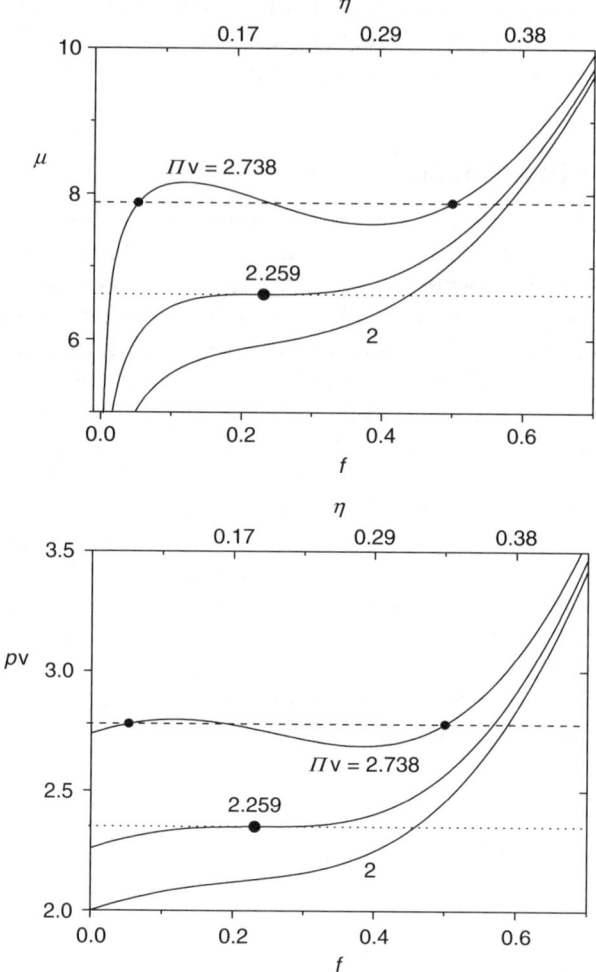

Figure 6.3 Colloid chemical potential μ_F and pressure $(pv)_F$ for a fluid with $q = 0.61$ as a function of f, for three values of Πv: 2 (lower curves, no GL phase separation), 2.259 (middle curves, critical conditions) and 2.738 (top curves, GL demixing with $f_G = 0.052$ and $f_L = 0.5$). In this example the Y terms in eqn (6.12) and (6.13) were neglected.

$q_s = 0.6$ ($q = 0.61$) for three values of Πv. For small Πv ($=2$ in Figure 6.3, bottom curves) both μ and pv increase monotonically with f (or η) so there can be no GL phase separation (there is FS coexistence, but that cannot be seen in this graph). For high Πv ($=2.738$ in Figure 6.3, top curves) we see a van der Waals loop in both μ and pv; the horizontal dashed lines at $\mu = 7.878$ and $pv = 2.779$ intersect the curves at $f_L = 0.5$ and $f_G = 0.052$ in both diagrams, so there is GL coexistence at these colloid concentrations. At $\Pi v = 2.259$ (middle curves) both μ and pv have an inflection point with zero slope of the curves at

$f = 0.231$; at this point $\mu' = \mu'' = p' = p'' = 0$, which is the condition for the critical point: for Πv below 2.259 no GL demixing is possible.

We note that the value of q (0.61 in Figure 6.3) is slightly too high to justify the neglect of the Y terms in eqn (6.12) and (6.13): this value is somewhat outside the colloid limit. It is perfectly possible to include these Y terms and find more complicated expressions for μ and pv. Then the quantitative detail of Figure 6.3 will change. However, the qualitative picture remains the same, with stable liquid only above the critical point and no liquid below it. In the remainder of this chapter the full form of eqn (6.12) and (6.13) is used, which makes the results also applicable outside the colloid limit.

6.3 Results

Figure 6.4 shows a phase diagram $y(\eta)$ for $q = 0.8$. The solid part at the bottom is the FS binodal. For $y = 0$ we have hard-sphere coexistence at $\eta_F = 0.49$ and $\eta_S = 0.54$, as demonstrated in Figure 6.2. With increasing y the FS demixing gap widens somewhat, with η_F slightly decreasing and η_S slightly increasing. At a certain value $y^{tp} = 1.316$ the F branch jumps discontinuously from $\eta_L^{tp} = 0.468$ to $\eta_G^{tp} = 0.013$ at constant y; at this value $y = y^{tp}$ the S branch shows a discontinuity in the slope at $\eta_S^{tp} = 0.565$. This is the GLS triple point at $q = 0.8$, which is characterized by four parameters: y^{tp} and three coexisting compositions. For $y > y^{tp}$ there is only GS demixing, as shown by the solid binodals

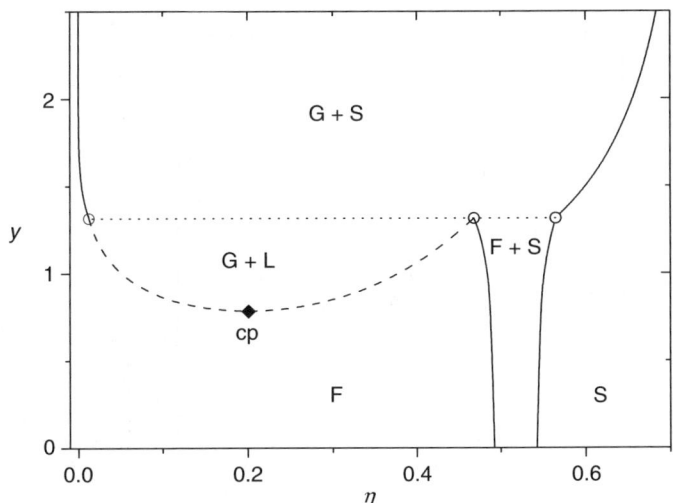

Figure 6.4 GL binodal (dashed line), GS binodal (solid line), and FS binodal (solid line) for $q = 0.8$. The diamond is the GL critical point at $y^{cp} = 0.782$, $\eta^{cp} = 0.201$. The horizontal dotted line is the triple point at $y^{tp} = 1.316$, $\eta_G^{tp} = 0.013$, $\eta_L^{tp} = 0.468$ and $\eta_S^{tp} = 0.565$. The demixing regions are indicated as G + S, G + L and F + S. Also the one-phase regions F and S are shown.

in the upper part of the phase diagram: a (very) dilute gas phase is in equilibrium with a (very) concentrated solid phase.

At the triple point ($y = y^{tp}$) there is GL demixing in the region $\eta_G^{tp} < \eta < \eta_L^{tp}$. For $y > y^{tp}$ no liquid is possible, but for $y < y^{tp}$ there exists stable liquid, over an η window which narrows as y decreases. At the critical point (marked as a diamond in Figure 6.4; $y^{cp} = 0.782$, $\eta^{cp} = 0.201$) this window is reduced to zero. Liquid is thus only stable over the range $y^{cp} < y < y^{tp}$, in the region indicated as G + L in Figure 6.4.

Figure 6.4 applies to one particular size ratio $q = R/a = 0.8$. Figure 6.5 shows how the picture changes when q is varied. This figure gives triple points (circles) and critical points (diamonds) for nine values of q: $q = 1$, 0.9, 0.8, 0.7, 0.6, 0.5, 0.45, 0.4 and 0.388; the latter value corresponds to the critical endpoint (see eqn (6.15)). For $q = 1$, 0.8, 0.6 and 0.45 the GL binodals (dashed lines) and FS binodals (solid lines) are indicated; in order not to overcrowd the figure the GS binodals above the triple point are omitted. The curves second from top ($q = 0.8$) are the same as in Figure 6.4.

We see in Figure 6.5 that for high q the liquid window is relatively wide, both in terms of y ($1 < y < 1.8$ for $q = 1$) and in terms of η ($0 < \eta < 0.49$ for $y = y^{tp} = 1.8$). As q decreases, this liquid window shifts to lower y (although initially the ratio y^{tp}/y^{cp} does not change much) and the η range becomes narrower (although for q above 0.6 the effect is small). When q drops below 0.5 the liquid window quickly shrinks, both in terms of y and in terms of η.

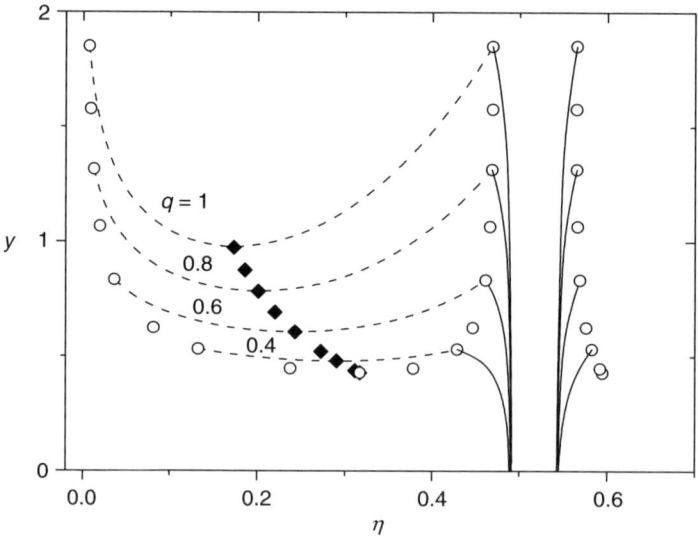

Figure 6.5 GL binodals (dashed lines) and FS binodals (solid lines) for $q = 1$, 0.8, 0.6 and 0.45. Triple points (circles) and critical points (diamonds) are indicated for $q = 1$, 0.9, 0.8, 0.7, 0.6, 0.5, 0.45, 0.4 and 0.388. The latter value corresponds to the *critical endpoint*, where the critical point coincides with (the fluid part of) the triple point.

At $q = q^{cep} = 0.388$ the liquid window and the GL binodal collapse into a single point: this is the *critical endpoint* (cep), where the critical point coincides with the triple point (which then actually is a double point as the G and L phases merge into one critical F phase which coexists with a solid). For $q < q^{cep}$ GL demixing is metastable: FS coexistence is then the stable situation.

The four coordinates of the cep follow analytically from the four equations $\mu' = \mu'' = 0$ (*i.e.* critical conditions) plus equal μ and pv in the F and S phases. The result is

$$q^{cep} = 0.388; \quad Y^{cep} = 1.466(y^{cep} = 0.428); \quad \eta_F^{cep} = 0.317; \quad (6.15)$$
$$\eta_S^{cep} = 0.594$$

This cep is a key feature of the phase diagram, as it determines the boundary condition for the existence of a stable liquid. It may be interpreted also in a quite different way, *i.e.* in terms of the *range* and the *strength* of the interaction between two colloids in a solution of nonadsorbing polymer. I do not go into any detail but mention only that the cep corresponds to the situation that the range is about one-third of the particle radius (*i.e.* $q_s^{cep} = 1/3$ in our present model) and the strength (*i.e.* the contact potential) is about $2kT$.[29] Liquid is only possible when the relative range is greater than $1/3$ and the strength weaker than $2kT$. For a shorter range and/or higher strength a crystalline solid is the preferred thermodynamic state for the condensed phase. This conclusion seems to be valid[31] not only for the present free-volume model, but also for quite different systems like a fluid of a Lennard-Jones type[32] and a Yukawa fluid.[33]

Figures 6.4 and 6.5 show phase diagrams in terms of the normalized polymer concentration $y = \varphi/\varphi^*$. This is the usual presentation reported in the literature. However, in Section 6.2.1 it was found that, especially for high q (protein limit) it is more convenient to use the parameter $Y = yq^{-1.3}$. It is useful to present Figure 6.5 also in the Y form. This is done in Figure 6.6, where now the q range is extended towards the protein limit: apart from the values given in Figure 6.5 also triple and critical points are shown for $q = 1.2, 1.4, 1.6, 1.8, 2, 3, 4$ and 5. The four GL and FS binodals are the same as in Figure 6.5.

Although the dataset in Figure 6.6 is identical to that in Figure 6.5 (at least for $q = 1$ and below), there are conspicuous differences between the two figures. These are related to the different behaviour of the triple points $Y^{tp}(q)$ as compared to $y^{tp}(q)$ and of the critical points $Y^{cp}(q)$ as compared to $y^{cp}(q)$.

In the triple point y^{tp} diverges as $q^{1.3}$ for high q. When y^{tp} is normalized to $Y^{tp} = y^{tp}q^{-1.3}$ we see a slight variation of Y^{tp} with q in the colloid limit: the minimum value, at the cep, is $Y^{cep} = 1.47$ at $q^{cep} = 0.39$, and Y^{tp} increases weakly with increasing q. However, for $q = 2$ or above Y^{tp} becomes essentially constant at a value $Y^{tp} = 2.1$. The variation of Y^{tp} over the entire range $q > q^{cep}$ is thus rather small: the protein-limit value is only 40% higher than the cep value.

For the critical point we see a qualitatively different behaviour for $Y^{cp}(q)$ as compared to $y^{cp}(q)$: starting from the cep, y^{cp} in Figure 6.5 goes up with q, whereas Y^{cp} in Figure 6.6 goes down. The reason is, obviously, the factor $q^{-1.3}$

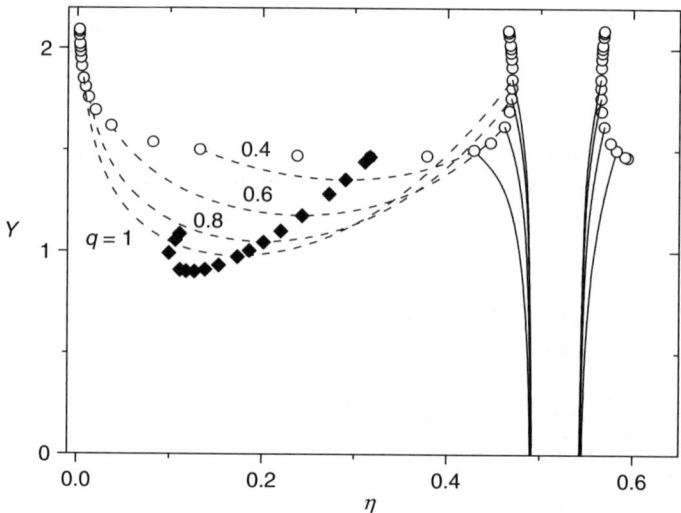

Figure 6.6 As Figure 6.5 but now with $Y = yq^{-1.3}$ along the ordinate axis. Triple and critical points (symbols) are given not only for the q set as in Figure 6.5 but, in addition, also for $q = 1.2$, 1.4, 1.6, 1.8, 2, 3, 4 and 5.

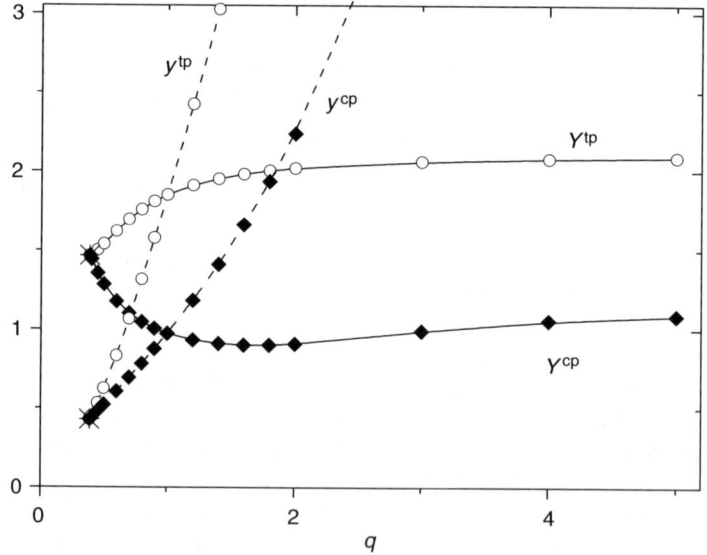

Figure 6.7 Data for triple points (circles) and critical points (diamonds) from Figures 6.5 and 6.6 in the form y^{tp} and y^{cp} (dashed curves) and Y^{tp} and Y^{cp} (solid curves) as a function of q. For Y the entire q set (symbols) is shown, from $q = q^{cep}$ to $q = 5$. For y^{tp} the highest q value shown is for $q = 1.2$, and for y^{cp} this is $q = 2$.

In the conversion from y to Y, which overcompensates the increase of y^{cp} with q in Figure 6.5. Also here we reach a more or less constant level ($Y^{cp} \approx 1$) in the protein limit, and for smaller q the variation of Y^{cp} with q is weak, as the cep value is $Y^{cp} = 1.47$.

Finally, I present the data for triple and critical points from Figures 6.5 and 6.6 in a different way in Figure 6.7: y^{tp} and y^{cp} (dashed curves), and Y^{tp} and y^{cp} (solid curves) as a function of q. This figure illustrates the width of the liquid window in terms of the polymer concentration. The asterisks in Figure 6.7 correspond to the cep, where the width of the liquid window is zero in terms of the *difference* $Y^{tp} - Y^{cp}$, or unity in terms of the *ratio* Y^{tp}/Y^{cp}. The maximum width, in the protein limit, is $Y^{tp} - Y^{cp} = 2$, and also $Y^{tp}/Y^{cp} = 2$. For the nonnormalized polymer concentrations we find $y^{tp} - y^{cp} = 2q^{1.3}$ (so this width diverges in the protein limit) and again $y^{tp}/y^{cp} = 2$. The liquid window is thus rather narrow, covering at most a factor of 2 in (external) polymer concentration.

6.4   Concluding Remarks

In this contribution given on the occasion of Brian's retirement I have described some features of the recent progress in theoretical understanding of polymer/colloid phase diagrams. Necessarily, I omitted most of the (quite interesting) details. For example, I discussed only phase diagrams in terms of the external concentration φ (or $y = \varphi/\varphi^*$). That is a transparent concept in the theoretical model, but it is not of immediate interest for an experimentalist, who has only access to internal concentrations $\phi = \alpha\varphi$. It is straightforward to calculate $\phi(\eta)$ diagrams for different q, as all the ingredients are available (see eqn (6.7) and (6.8) for α). However, a full description would require a much longer chapter. Elsewhere, Remco Tuinier and myself will present[34] a more elaborate treatment. Also simple explicit analytical approximations for the various parameters such as y, Y, φ, ϕ, η in triple and critical points and along the binodals as a function of q are now available.

Another important aspect which I did not address here is how our new theory compares with experimental data and with other theoretical treatments, including simulations. Let me say only this: as far as we can tell now we find rather good agreement with experiment, not only in the colloid limit (where such agreement was established before) but also for higher q and the protein limit (where, so far, large discrepancies have existed). As to simulations, our $y \sim q^{1.3}$ scaling in the protein limit is nicely reproduced,[23] and other theories[35,36] seem to follow this scaling law as well (although the numerical prefactors may differ somewhat).

Acknowledgements

First of all, I am very grateful to Brian himself. We collaborated regularly for many years on depletion and other things, which culminated in our book

Polymers at Interfaces, with Martien Cohen Stuart, the late Jan Scheutjens and myself from Wageningen and Brian Vincent and Terence Cosgrove from Bristol. Over the last six years Brian and myself were also in regular contact in the context of the International Association of Colloid and Interface Scientists (IACIS): Brian as president-elect and then as president, myself as honorary secretary and treasurer. I have served under several presidents: all were good but Brian was best!

With respect to the present chapter, I am greatly indebted to my former PhD student Remco Tuinier (Jülich), who put me on the track of the Lekkerkerker theory and with whom I have had intense cooperation over the last five years. Remco was so kind to prepare the figures for this chapter.

Finally, I should mention Sasha Skvortsov (St Petersburg), who rediscovered a half-forgotten equation in our book that we later translated as $\delta^{-2} = \delta_0^{-2} + \xi^{-2}$, which is the basis for eqn (6.12) in this chapter; also he played an important role in the additivity rule $\Pi = \Pi_0 + \Pi_{sd}$ which leads to eqn (6.13).

References

1. J.A. Long, D.W.J. Osmond and B. Vincent, *J. Colloid Interface Sci.*, 1973, **42**, 545.
2. C. Cowell, R. Li-In-On and B. Vincent, *J. Chem. Soc., Faraday Trans. 1*, 1978, **74**, 337.
3. B. Vincent, *Croatica Chimica Acta*, 1983, **56**, 615.
4. B. Vincent, in *Future Directions in Polymer Colloids*, ed. M. El-Aasser and R.M. Fitch, Martinus-Nijhoff, 1987. p. 191.
5. B. Vincent, J. Edwards, S. Emmett and A. Jones, *Colloids Surf.*, 1986, **17**, 261.
6. S. Rawson, K. Ryan and B. Vincent, *Colloids Surf.*, 1988, **34**, 89.
7. A. Jones and B. Vincent, *Colloids Surf.*, 1989, **42**, 113.
8. A. Milling, B. Vincent, S. Emmett and A. Jones, *Colloids Surf.*, 1991, **57**, 185.
9. P. Jenkins and B. Vincent, *Langmuir*, 1996, **12**, 3107.
10. A.P. Gast, W.B. Russel and C.K. Hall, *J. Colloid Interface Sci.*, 1986, **109**, 161.
11. F. Leal-Calderon, J. Bibette and J. Biais, *Europhys. Lett.*, 1993, **23**, 653.
12. S.M. Ilett, A. Orrock, W.C.K. Poon and P.N. Pusey, *Phys. Rev. E*, 1995, **51**, 1344.
13. J. Edwards, D.H. Everett, T. O'Sullivan, I. Pangalou and B. Vincent, *J. Chem. Soc., Faraday Trans. 1*, 1984, **80**, 2599.
14. B. Vincent, *Chem. Eng. Sci.*, 1987, **47**, 779.
15. B. Vincent, J. Edwards, S. Emmett and R. Croot, *Colloids Surf.*, 1988, **31**, 267.
16. S. Emmett and B. Vincent, *Phase Transitions*, 1990, **21**, 197.
17. G.J. Fleer, M.A. Cohen Stuart, J.M.H.M. Scheutjens, T. Cosgrove and B. Vincent, *Polymers at Interfaces*, Chapman & Hall, London, 1993.

18. H.N.W. Lekkerkerker, *Colloids Surf.*, 1990, **51**, 419.
19. H.N.W. Lekkerkerker, W.C.K. Poon, P.N. Pusey, A. Stroobants and P.B. Warren, *Europhys. Lett.*, 1992, **20**, 559.
20. P.G. de Gennes, *Scaling Concepts in Polymer Physics*, Cornell University Press, Ithaca, NY, 1979.
21. P.N. Pusey and W. van Megen, *Nature*, 1986, **320**, 340.
22. W.G. Hoover and F.M. Ree, *J. Chem. Phys.*, 1968, **49**, 3609.
23. G.J. Fleer and R. Tuinier, *Phys. Rev. E*, in press.
24. D.G.A.L. Aarts, R. Tuinier and H.N.W. Lekkerkerker, *J. Phys.: Condens. Matter*, 2002, **14**, 7551.
25. N.F. Carnahan and K.E. Starling, *J. Chem. Phys.*, 1969, **51**, 635.
26. C.K. Hall, *J. Chem. Phys.*, 1972, **52**, 2252.
27. G.J. Fleer, A.M. Skvortsov and R. Tuinier, *Macromol. Theory Simul.*, 2007, **16**, 531.
28. A. Hanke, E. Eisenriegler and S. Dietrich, *Phys. Rev. E*, 1999, **59**, 6853.
29. A.A. Louis, P.G. Bolhuis and E.J. Meijer, *J. Chem. Phys.*, 2002, **116**, 10547.
30. L. Schäfer, *Excluded-Volume Effects in Polymer Solutions*, Springer Verlag, Berlin, 1999.
31. G.J. Fleer and R. Tuinier, *Physica A*, 2007, **379**, 52.
32. G.J. Vliegenhart, J.F. Lodge and H.N.W. Lekkerkerker, *Physica A*, 1999, **263**, 378.
33. R. Tuinier and G.J. Fleer, *J. Phys. Chem. B*, 2006, **110**, 20540.
34. G.J. Fleer and R. Tuinier, in preparation.
35. A. Pelissetto and J.P. Hansen, *Macromolecules*, 2006, **39**, 9571.
36. M. Fuchs and K.S. Schweizer, *J. Phys.: Condens. Matter*, 2002, **14**, R239.

Chapter 7

Nanobubbles, Dissolved Gas, Boundary Layers and Related Mysterious Effects in Colloid Stability

John Ralston

IAN WARK RESEARCH INSTITUTE, UNIVERSITY OF SOUTH AUSTRALIA, MAWSON LAKES CAMPUS, MAWSON LAKES, ADELAIDE, SA 5095, AUSTRALIA

Abstract

My scientific and personal relationship with Brian Vincent commenced in 1978 when I came to Bristol to work on static and dynamic light scattering studies of microemulsions. I worked with Ron Ottewill at Bristol, Peter Pusey at the RSRE in Malvern and Tharwat Tadros at ICI in Jealott's Hill. During this period of six months at Bristol, I got to know Brian and very much appreciated his studies on colloid stability, steric stabilisation, the nature of adsorbed layers and the like. Over the period since 1978, our scientific interactions and friendship have blossomed tremendously. We have exchanged PhD students and some excellent postdoctoral fellows. My research interests are aimed at answering questions about why bubbles and particles interact, the static and dynamic aspects of wetting and colloid stability. In this chapter, I review some of our most recent work on the way in which dissolved gas and surface nanobubbles influence colloid stability. Along the way I illustrate how these "boundary layer" bubbles and adsorbed gas layers play an important role, drawing on many of the concepts that Brian has developed and used over the many important years of his scientific research.[1,2]

New Frontiers in Colloid Science: A Celebration of the Career of Brian Vincent
Edited by Simon Biggs, Terence Cosgrove and Peter Dowding

7.1 Introduction

Bubble formation at solid–liquid interfaces has long been of interest. Robert Boyle's[3] observations of a bubble in the eye of a viper, which had been placed in an evacuated chamber, are perhaps the earliest reported record. It has been convincingly demonstrated that the presence of nuclei in solution or on solid surfaces is a prerequisite for bubble formation.[4–7] Dissolved gas can exist in molecular form, as a "solution" in the form of tiny bubbles, purportedly down to nanometre dimensions, "bubstons",[8] "gas bells"[9] and as somewhat larger bubbles. The formation of larger bubbles from preexisting sources of bubbles can occur without a nucleation step.[9] Small bubbles with a small Peclet number (submicrometre in diameter), Harvey nuclei and entrained or sparged bubbles can act as a source. In the specific case of Harvey nuclei, the source can be any gas-filled re-entrant cavity, where the trapped gas cannot be displaced by the surrounding liquid phase.[10]

These tiny bubbles, existing as a dispersion or as Harvey nuclei resident on surfaces, may be removed by high-speed centrifuging and subjection to high pressures (above 700 atm) or boiling followed by exposure to high pressures.[7,11] Gas nuclei have been found to be more difficult to remove from a rough and irregular hydrophobic surface than from any other type of surface.[9,12] These results are in accord with extant theory of bubble nucleation and detachment processes.[10]

Interparticle forces in a colloidal suspension can often be described by classical DLVO theory; however, in systems where the particle surface is hydrophobic, an additional non-DLVO force is often introduced. This is an attractive force, with some studies linking its range and magnitude to the concentration of dissolved gas.[13–15] Different theories exist to explain the origin of this hydrophobic attraction and, in particular, it has been proposed that these attractive forces are due to the presence of very small bubbles on the particle surface.[16–18] The presence of small gas bubbles on hydrophobic surfaces was demonstrated indirectly in the key early experiments of Harvey *et al.*[19–21] and observed on hydrophobic silica, glass, mica and graphite surfaces using tapping-mode atomic force microscopy (TMAFM).[22–25] It appears that both the physical and chemical natures of the solid surface influence the hydrophobic force, by promoting or suppressing the formation of very small surface bubbles.[26]

Regardless of the source of these very small surface bubbles, it is evident that the reported range of hydrophobic attraction varies greatly in surface force studies, ranging from those that match DLVO theory[16,26] to studies that show long-range forces extending up to 250 nm.[28] A surface covered by very small bubbles will by nature be heterogeneous, displaying regions with different physical and chemical properties. Therefore, since only a small portion of the surfaces can interact in these experiments, a random distribution of surface bubbles may explain some of the variation observed in the measured ranges of hydrophobic attraction.

Before the advent of modern force-measurement techniques, traditional studies of aggregation rates and stability ratios allowed the indirect assessment of interparticle forces. For studies of heterogeneous surfaces, there are obvious statistical advantages in dealing with a large ensemble of particles, particularly when bulk properties like colloid stability are to be predicted. An additional advantage of the aggregation rate approach over studies that employ fixed surfaces is the ability of the particles to rotate freely during collisions, a factor which may prove important for heterogeneous surfaces. However, despite these advantages, recent studies of this type on colloidal systems are very rare. This may partly be due to systems where classical DLVO theory fails to predict accurately the dependence of colloid stability on electrolyte concentration.[29] Recent studies have shown, however, that when the surface charge of the particles is sufficiently low, classical DLVO theory adequately predicts the stability/electrolyte dependence.[30] In the case of silica, the surface charge is both low and well defined.[31]

Empirical, hydrophobic attractions are usually described using a single- or double-exponential term where one or two decay lengths are invoked.[26] Recently, our theoretical analysis has shown that the presence of very small bubbles on the surface of a particle can lead to an attraction that appears similar to the action of hydrophobic forces. The origin of the effect lies in a change in the magnitude (and even sign) of the van der Waals interactions between a particle and a macrobubble.

Our first objective[27] was to investigate the effects of surface hydrophobicity and heterogeneity, together with dissolved gas concentration, on the stability of a colloidal particle suspension. On the basis of the premise that very small bubbles attached to particle surfaces influence hydrophobic attraction, conditions are varied to suppress or promote the formation of bubbles. The results are modelled using classical DLVO theory, introducing the influence of gas at the particle–solution interface.

Recently, with the advent of TMAFM, imaging of nanometric or sub-micrometric bubbles formed on solid surfaces has become of considerable interest to colloid science.[33–35] This research was spurred by a desire to explain the nature of the long-range hydrophobic forces often observed between hydrophobic solid surfaces immersed in aqueous solutions.[36]

The TMAFM technique can "image" very small bubbles on solid substrate surfaces, since it has the capacity to determine both the physical structure and physicochemical nature of soft and fragile surfaces at the molecular level, provided that these surfaces are stable during scanning.

The results of the TMAFM imaging of very small bubbles have yielded quite diverse interpretations for "nanobubble" formation at solid–water interfaces. Lou *et al.*[33] showed that small bubbles less than 100 nm in diameter could form on both hydrophobic (highly oriented pyrolytic graphite) and hydrophilic (freshly cleaved mica) surfaces. Using silicon wafer substrates modified by octadecyltrichlorosilane (OTS) (root mean square (RMS) roughness <0.2 nm; water advancing contact angle (θ_w) of 110 °), Ishida *et al.*[34] observed that small bubbles with a base diameter of 650 nm and a height of 40 nm were *linearly* distributed on the substrate surface of observation. Accompanying AFM force measurements suggested the existence of a long-range hydrophobic force between surfaces with bubble domains. TMAFM imaging by Tyrrell and Attard[35] showed domains which were very different from those observed by Ishida *et al.* The bubble domains cover the whole hydrophobised surface of the substrate (methylated by dichlorodimethylsilane vapour; RMS roughness <0.5 nm; $\theta_{water,advancing} = 101°$, $\theta_{water,receding} = 80°$, hysteresis of $21°$). The domains had a mean height of 20–30 nm and mean diameter of 71–87 nm. Since the surface roughness of the substrate surfaces used in these two studies is rather similar, the differences in behaviour are not readily explained.

Our second objective[36] was to investigate the formation of very small bubbles at solid–water interfaces through TMAFM imaging of well-defined substrate surfaces with known degrees of surface roughness, hydrophobicity and gas supersaturation. We demonstrate that there is a significant difference between the macroscopic contact angle for these substrates and the microscopic angle when a very small bubble is formed. The origin of this difference may be due to the line tension acting at the solid–water–vapour contact line.

Our third objective[37] was to study the rate at which gas was adsorbed and nanobubbles formed. There have been various investigations of the kinetics of bubble formation.[13,38–41] In particular, Carr *et al.*[40] investigated the kinetics of nucleation of H_2 bubbles on an electrode surface, using the quartz crystal microbalance (QCM) technique. This technique has been used to study the adsorption of organic molecules on solid surfaces.[42–44] In the QCM measurement, the mass of adsorbed molecules is evaluated continuously as a change in frequency with time. Due to the high sensitivity and fast response of the QCM, we aimed to probe whether the early stage(s) of gas bubble nucleation at water–solid interfaces could be investigated, before TMAFM imaging is feasible.

Thus, the adsorption of CO_2 molecules onto a QCM coated gold electrode from aqueous solution was monitored as a function of time and hydrophobicity of the electrode surface under the same experimental conditions used in our TMAFM and colloid stability experiments.

7.2 Results and Discussion

7.2.1 Stability Ratio as a Function of Dissolved Gas and Hydrophobicity

The stability ratio (W) as a function of salt concentration (C) is shown in Figure 7.1. The most important features of the data in Figure 7.1 are the variation between the curves for the particles with hydrophilic and hydrophobic surfaces and the response of these systems to increased concentrations of dissolved gas.

In the case of the hydrophilic and dehydroxylated particles, it can be clearly seen that there is no detectable difference between the stability of a suspension under normal conditions compared with that of an identical system with an increased concentration of dissolved gas. Smooth hydrophilic surfaces will wet completely on immersion in water and are not expected to nucleate very small

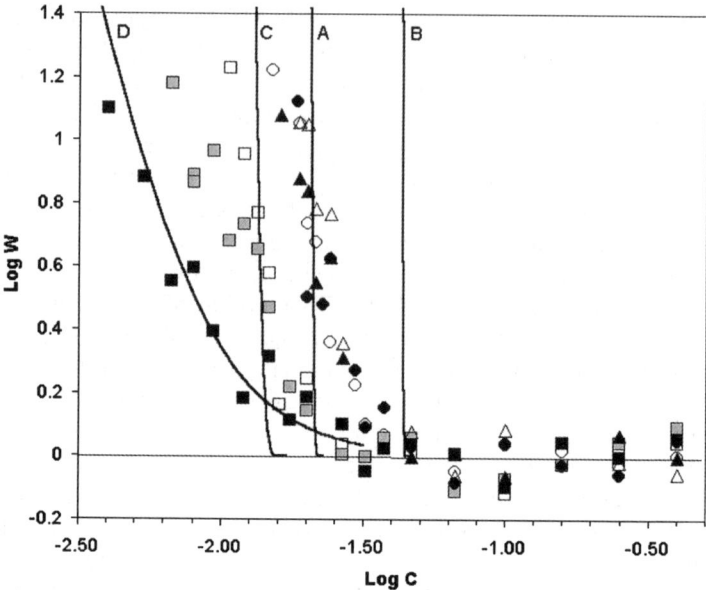

Figure 7.1 Stability ratio *versus* KCl concentration for Geltech silica spheres at pH = 4.2 with a clean hydrophilic surface under normal conditions (○), clean hydrophilic surface/dissolved CO_2 concentration 10^{-2} M (●), methylated hydrophobic surface/normal conditions (▨), methylated hydrophobic surface/dissolved CO_2 concentration 10^{-2} M (■) and methylated hydrophobic surface/degassed (□), dehydroxylated surface/normal conditions (△), dehydroxylated surface/dissolved CO_2 concentration 10^{-2} M (▲). Solid lines, A, B, C and D indicate calculated stability curves. (Reproduced with permission from the American Chemical Society, *Journal of Physical Chemistry B*, 2003, **107**, 13, Figure 3.)

surface bubbles under supersaturated conditions to the extent that there is any detectable impact on colloid stability. In the case of smooth dehydroxylated but hydrophobic particles where the advancing contact angle is 41°, dissolved gas does not influence the dispersion stability. This is in accord with prior colloid probe determinations of the interaction force between dehydroxylated silica surfaces where the "jump distances" in the presence of gas-saturated (air, argon and CO_2) solutions were generally found to be less than a few nanometres. Similar conclusions have been reached for smooth hydrophobic ($\theta_{ad} < 38°$) alumina surfaces.

Conversely, there is a clear link between the level of dissolved gas and stability of the suspensions where the particles have a methylated surface. First, there is a marked decrease in the gradient of the stability curve (Figure 7.1) as the dissolved gas concentration increases. There is an increased attraction between the particles compared with the clean and dehydroxylated cases. This behaviour is in accord with an earlier *in situ* Fourier transform infrared study of the aggregation behaviour of smooth (a) dehydroxylated and (b) methylated

0.5 µm diameter Geltech silica spheres at 10^{-3} M KNO_3 in the presence of carbon dioxide.[45] A second feature of note in Figure 7.1 is the displacement of the stability curves for the methylated silica spheres to lower critical coagulation concentration (ccc) values compared with the clean and dehydroxylated cases. We also remark that the gradient of the log W *versus* log C plot for methylated particles under degassed conditions matches the gradient for the clean, hydrophilic particles.

To establish whether or not the presence of dissolved gas influences the electrical double-layer characteristics of the particle surfaces, the zeta potential of the silica particles (clean, dehydroxylated and methylated) was determined at pH = 4.2 and 10^{-3} M KCl at various dissolved gas concentrations. The average zeta potential data, representing three separate experiments, are given in Table 7.1.

For the clean, hydrophilic particles, there is no influence of dissolved gas, within experimental error. In the case of dehydroxylated particles, again there is no influence of dissolved gas under normal and elevated CO_2 conditions, with a slight increase only in magnitude when the system is degassed. The reason for the significant difference in zeta potential between these two silica samples lies in the dehydroxylation process. The latter removes surface charging sites, where charge can develop, as silanol groups are converted to siloxane bridges.[46] In the case of the methylated surfaces, there is a marked influence of dissolved gas concentration on zeta potential when the suspensions are degassed.

We have observed similar behaviour in a related study, where nitrogen-saturated solutions of aqueous electrolyte solution, at elevated pressures, were forced through methylated silica capillaries in streaming-potential investigations. In this study, the marked (up to 50%) change in the magnitude of the zeta potential was ascribed to the presence of very small bubbles attached to the rough (on the nanometre scale), heterogeneous (patchy) methylated silica surfaces. These bubbles could be stripped from the capillary surfaces under conditions of very high shear, or prevented from forming by rendering the surface hydrophilic with a non-ionic surfactant.

Table 7.1 Average zeta potential of Geltech silica spheres at pH = 4.2 and KCl concentration of 10^{-3} M as a function of dissolved gas concentration.

Particle type/conditions	Average zeta potential ± 95% confidence limits (mV)
Clean/degassed	-29.9 ± 1.2
Clean/normal	-31.4 ± 1.0
Clean/gassed, CO_2 10^{-2} M	-30.0 ± 1.3
Dehydroxylated/degassed	-21.4 ± 1.1
Dehydroxylated/normal	-16.7 ± 0.7
Dehydroxylated/gassed, CO_2 10^{-2} M	-17.6 ± 1.6
Methylated/degassed	-36.6 ± 3.6
Methylated/normal	-16.4 ± 4.5
Methylated/gassed, CO_2 10^{-2} M	-22.2 ± 3.7

7.2.2 Colloid Stability Analysis

Classically, the forces experienced between two like particles in suspension can be described as the sum of the attractive van der Waals forces and the repulsive electrostatic forces. In terms of interaction energies this can be expressed as a function of H, the shortest distance between the particle surfaces:

$$V(H) = V_R(H) + V_A(H) \tag{7.1}$$

There are various equations describing the electrostatic repulsion as a function of distance. The constant charge model for identical particles shown in eqn (7.2) was chosen since it was best able to model the experimental data. AFM colloid probe studies using similar silica spheres have also found that the constant charge model gives the best fit to experimental data. The approximations used for deriving eqn (7.2) restrict its use to low surface potentials:

$$V_R(H) = -\frac{64\pi a n k_B T}{\kappa^2} \gamma^2 \ln[1 - \exp(-\kappa H)] \tag{7.2}$$

Where

$$\gamma = \tanh \frac{z e \Psi_d}{4 k_B T} \tag{7.3}$$

In regard to eqn (7.2) and (7.3), k_B is the Boltzmann constant, T is the temperature in kelvin, z is the valency of the ions (assuming a symmetric electrolyte), e is the charge of one electron, Ψ_d is the potential at the outer Helmholtz plane, a signifies the radius of the particle, n is the bulk number density of ions and κ is the Debye–Hückel parameter with units of reciprocal length. It is generally assumed that Ψ_d and zeta potential are similar.

When considering the attractive van der Waals forces at short distances ($H \ll a$), the interaction energy between identical particles can be adequately described by[47]

$$V_A = -\frac{Aa}{12H} \tag{7.4}$$

where A refers to the Hamaker constant for two identical particles immersed in a bathing medium.

Using these expressions, it is therefore possible to calculate the theoretical stability ratio W at any given salt concentration:

$$W = 2 \int_2^\infty \exp\left(\frac{V(s)}{k_B T}\right) \frac{ds}{s^2} \tag{7.5}$$

where $s = R/a$ and R denotes the distance between the centres of the approaching spheres. This stability ratio was later redefined by McGown and

Parfitt[48] taking into consideration the fact that the attractive van der Waals force still operates during rapid coagulation:

$$W = \frac{\int_2^\infty \exp\frac{V(s)/k_BT}{s^2}\,ds}{\int_2^\infty \exp\frac{V_A(s)/k_BT}{s^2}\,ds} \tag{7.6}$$

This equation can be further refined by including the factor β, proposed by Honig *et al.*,[49] to correct for the influence of hydrodynamic interactions, which were shown to be of importance in the work of Spielman:[50]

$$W = \frac{\int_2^\infty \beta(s)\exp\frac{V(s)/k_BT}{s^2}\,ds}{\int_2^\infty \beta(s)\exp\frac{V_A(s)/k_BT}{s^2}\,ds} \tag{7.7}$$

where

$$\beta = \frac{6s^2 - 11s}{6s^2 - 20s + 16} \tag{7.8}$$

Since a theoretical plot of the stability ratio *versus* salt concentration can be obtained, a comparison of this theoretical plot in the slow regime below the ccc with the experimental version can be performed. By using both the Hamaker constant and surface potential as fitting parameters, the calculated stability curve can be matched to the experimental results. This has been performed recently using these equations by Puertas and Nieves[51] in a study of the stability of latex colloids. Although the accuracy of the Hamaker constant and surface potential obtained by this dual-parameter technique can be debated, the shape of the calculated curve does correspond very closely to the experimental curve in this particular study. For the stability curves presented in this chapter however, the particle surface potentials have been obtained from experimental results.

Using electrophoresis techniques, zeta potentials at specific salt concentrations may readily be obtained through experiments performed at KCl concentrations up to 10^{-2} M. At higher salt concentration, electrolysis becomes an issue. From a of a plot of ln(zeta potential) *versus* κ, the zeta potential near the ccc can be accessed by simple extrapolation under conditions where experimental measurements of mobility are difficult.[27]

With respect to the stability curves, a sensitivity analysis performed on the theoretical stability curve reveals several important points. An increase in the Hamaker constant shifts the stability curve to lower salt concentrations, but has little influence on the gradient, whereas a decrease in particle radius decreases the gradient of the curve whilst the ccc remains relatively constant. It is also found that any variation in the zeta potential changes both the position of the curve and, to a small degree, the gradient. This sensitivity analysis is valuable for it provides guidance in the following data interpretation.

A theoretical stability curve was determined for the clean hydrophilic particles (Figure 7.1, curve A). The Hamaker constant used in the calculation was adjusted to obtain the best fit with the experimental data.[27] It is notable that the agreement between the experiment and theory for the case of hydrophilic or dehydroxylated silica is acceptable, in line with recent studies of the stability of colloidal dispersions of low surface charge. The presence of *very* short range hydration forces has been detected between silica surfaces, irrespective of whether the latter are hydrophilic or dehydroxylated.[32] Under the conditions of this present study, they do not influence the essential coagulation behaviour, since regimes of both slow and fast aggregation are clearly evident as the salt concentration is increased.

We now address the case of methylated systems in the near-absence of dissolved gas and where gas is present. Before doing so, we remark that the evidence linking the presence of very small bubbles with surface heterogeneity (physical and/or chemical) is very strong indeed.[33-35] There are three reported TMAFM observations of "nanobubbles" on surfaces immersed in aqueous solutions reported to date, involving silane-treated silica or glass surfaces treated with silanes, cleaved mica and pyrolitic graphite. In only one of these studies was any attempt made to quantify the surface heterogeneity through the rudimentary determination of contact angle hysteresis (substantial, at 21°, compared with fluoropolymer surfaces). In view of the zeta potential data in Table 7.1, and the observed heterogeneous nature of our methylated surfaces, the presence of very small surface bubbles would appear to influence the observed stability behaviour. We have already demonstrated that various surface distributions of very small bubbles can alter the van der Waals interaction when a particle and a macrobubble interact. We proceed, knowing that we are dealing with a statistically large number of particles and therefore use average values for methylation and gas "layer thicknesses". In the case of the methylated particles, the methylation layer thickness was taken to be 0.2 nm, representing the difference between the RMS roughness data for the clean and methylated silica spheres. This value is consistent with the known chemistry of silanation processes for silica surfaces.

For the case of the hydrophobic methylated particles under degassed conditions (Figure 7.1, curve C), it was first assumed that the observed shift in the stability curve was due to the methylation layer altering the Hamaker constant of the surface, rather than to the presence of very small surface bubbles. To allow for this and any "air layer" present in later cases, a more complicated model incorporating two adsorbed layers was used for the calculation of the Hamaker constant. Using the generic approach of Usui and Barouch[52] as a basis, the following equation was obtained for the case of two interacting spheres of radius a_1 and a_2 with two adsorbed layers on each particle (Figure 7.2 describes the nomenclature used in eqn 7.10–7.27):

$$V_A = -\frac{1}{6}\left(\frac{a_1 a_2}{a_1 + a_2}\right)$$
$$\times \left(\frac{A_1}{D_1} + \frac{A_2}{D_2} + \frac{A_3}{D_3} + \frac{A_4}{D_4} + \frac{A_5}{D_5} + \frac{A_6}{D_6} + \frac{A_7}{D_7} + \frac{A_8}{D_8} + \frac{A_9}{D_9}\right) \qquad (7.9)$$

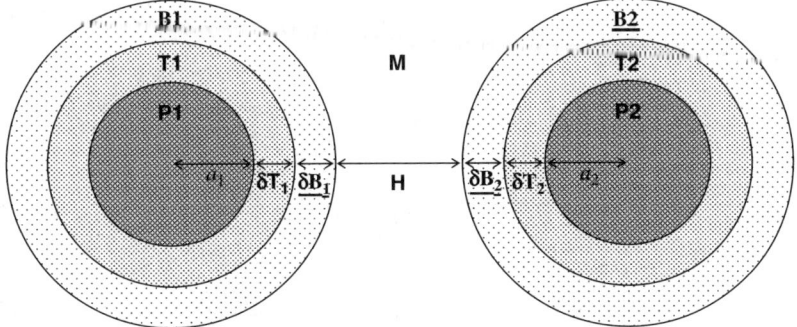

Figure 7.2 Arrangement of two identical spheres with a hydrophobic layer and air layer defining the nomenclature used in eqn (7.10)–(7.27).

where

$$A_1 = \left(A_{B1}^{1/2} - A_M^{1/2} \right) \left(A_{B2}^{1/2} - A_M^{1/2} \right) \tag{7.10}$$

$$D_1 = H \tag{7.11}$$

$$A_2 = \left(A_{T1}^{1/2} - A_{B1}^{1/2} \right) \left(A_{B2}^{1/2} - A_M^{1/2} \right) \tag{7.12}$$

$$D_2 = H + \delta B_1 \tag{7.13}$$

$$A_3 = \left(A_{P1}^{1/2} - A_{T1}^{1/2} \right) \left(A_{B2}^{1/2} - A_M^{1/2} \right) \tag{7.14}$$

$$D_3 = H + \delta T_1 + \delta B_1 \tag{7.15}$$

$$A_4 = \left(A_{B1}^{1/2} - A_M^{1/2} \right) \left(A_{T2}^{1/2} - A_{B2}^{1/2} \right) \tag{7.16}$$

$$D_4 = H + \delta B_2 \tag{7.17}$$

$$A_5 = \left(A_{T1}^{1/2} - A_{B1}^{1/2} \right) \left(A_{T2}^{1/2} - A_{B2}^{1/2} \right) \tag{7.18}$$

$$D_5 = H + \delta B_1 + \delta B_2 \tag{7.19}$$

$$A_6 = \left(A_{P1}^{1/2} - A_{T1}^{1/2}\right)\left(A_{T2}^{1/2} - A_{B2}^{1/2}\right) \tag{7.20}$$

$$D_6 = H + \delta T_1 + \delta B_1 + \delta B_2 \tag{7.21}$$

$$A_7 = \left(A_{B1}^{1/2} - A_M^{1/2}\right)\left(A_{P2}^{1/2} - A_{T2}^{1/2}\right) \tag{7.22}$$

$$D_7 = H + \delta B_2 + \delta T_2 \tag{7.23}$$

$$A_8 = \left(A_{T1}^{1/2} - A_{B1}^{1/2}\right)\left(A_{P2}^{1/2} - A_{T2}^{1/2}\right) \tag{7.24}$$

$$D_8 = H + \delta B_1 + \delta B_2 + \delta T_2 \tag{7.25}$$

$$A_9 = \left(A_{P1}^{1/2} - A_{T1}^{1/2}\right)\left(A_{P2}^{1/2} - A_{T2}^{1/2}\right) \tag{7.26}$$

$$D_9 = H + \delta T_1 + \delta B_1 + \delta T_2 + \delta B_2 \tag{7.27}$$

Using these equations, a single 0.3 nm organic layer (T1 and T2) with Hamaker constant of 4.0×10^{-20} J (with respect to vacuum[43]) was chosen to simulate the TMCS coating, since this value lies within the range reported for similar hydrocarbons. The calculated stability curve for these conditions however does *not* explain the experimental observations; rather the predicted displacement is to the right as a result of the lower van der Waals attraction (Figure 7.1, curve B).

Even under degassed conditions, some very small gas pockets are likely to remain trapped at surface sites. Although these sites do not constitute a uniform layer over the surface of the particle, to simplify the present calculation and allow their potential effect on the van der Waals forces to be determined, a second, outer air "layer" (B1 and B2) was introduced into the calculations. If this layer is 3 nm thick, then the corresponding calculated stability curve matches the experimental curve quite closely (Figure 7.1, curve C).

We note that the surface potential used in our model is obtained for an ensemble of particles. The description of the surface charge distribution on a surface decorated, to some degree, with very small bubbles is not tractable at present, and remains as a future problem. The recent approach of Zembala and Adamczyk[53] for particles on surfaces is very encouraging, however. If the concentration of surface bubbles increases, the net effect is to decrease the magnitude of the zeta and therefore the surface potential.[27] Although the exact potential of the gas–water interface remains elusive, studies indicate that it is likely to be rather small under the experimental conditions of this investigation. The observed change in the potential of the surface covered by very small bubbles is due to the underlying solid surface being screened by these bubbles.

Given that these surface bubbles increase their coverage of the methylated surface under normal and gassed conditions, it is helpful to consider the various possible surface interactions that may occur before attempting to describe the stability curves for these remaining cases. With regard to Figure 7.3 it is obvious that three distinct interactions can occur. Significantly, the bubble–water–silica van der Waals interaction is repulsive, whereas the bubble–water–bubble and silica–water–silica interactions are attractive, as shown in Figure 7.3.

In reality the total interaction energy between the particles will reflect a combination of these forces. The distribution of bubble coverage and diameters over the surface will further complicate any calculations. From the simplified interaction analysis, as the size and coverage of surface bubbles increase, one would expect that the bubble–water–bubble interactions would predominate. The AFM studies described below have shown that very small bubbles on methylated silica surfaces immersed in aqueous solutions with similar

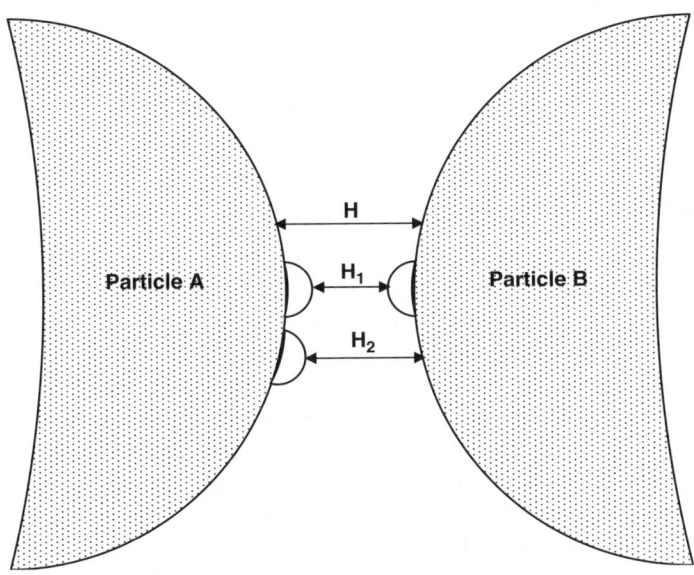

Figure 7.3   Scale diagram representing two 1 μm diameter particles with 100 nm diameter surface bubbles demonstrating the three separate interactions that can occur. In terms of the van der Waals forces, these are: strongly attractive bubble–water–bubble interactions occurring over distance H_1; strongly repulsive silica–water–bubble interactions occurring over distance H_2; and attractive silica–water–silica interactions occurring over distance H.

electrolyte and dissolved gas levels have radii between 20 and 60 nm and a surface coverage of approximately one-half. As a consequence, the effective radius of interaction between the particles will be that of the protruding surface bubbles rather than that of the particles themselves (see Figure 7.3).

Acknowledging the problems associated with the precise determination of the Hamaker constant in this situation, we have used the following procedure to calculate curve D in Figure 7.1. The effective interaction radius was taken as 40 nm on the basis of extant AFM evidence. Since the surface coverage was not known exactly, it was not possible to calculate a composite Hamaker constant with acceptable reliability by following the procedure we have used previously. Hence, we have used the Hamaker constant as an effective fitting parameter while still retaining all of the other features used to calculate curve C in Figure 7.1. In practice, this means that the outermost layer is taken as a composite air/water region with a thickness of 3 nm.[27] After the fitting exercise, the effective Hamaker constant for the layer B_1 ($= B_2$ for identical particles, Figure 7.2) is obtained.

Most significantly, the stability curve calculated for interacting surfaces decorated with protruding bubbles with a radius of 40 nm fits the gradient of the experimental data very closely (Figure 7.1, curve D). The change in the effective interaction radius caused by the protruding bubbles explains the change in the gradient of the experimental stability curves for the methylated particles as dissolved gas levels are increased (Figure 7.1). Furthermore, sensitivity analysis reveals that these changes in the gradient, which is the main feature of the data, cannot be explained by changes in the effective Hamaker constant because the latter does not alter the gradient.

7.2.3   TMAFM Imaging

Imaging of substrate surfaces immersed in CO_2 saturated aqueous solutions. The various wafer substrate surfaces were examined by TMAFM when immersed in dissolved CO_2 Milli-Q water solution (CO_2: 0.01 M).

Imaging of clean hydrophilic silicon wafer surfaces. A TMAFM image of hydrophilic silicon wafer substrates immersed in the CO_2 solution evidenced a smooth substrate surface and is very similar to that obtained in air.

Dehydroxylated silicon wafer surface. As in the case of hydrophilic substrates, the image obtained showed smooth surface features, again similar to those in air. These results are consistent with our high-energy reflectivity measurements of smooth hydrophobic solid–water interfaces.

TMCS vapour-prepared silicon wafer surface. TMAFM height images for TMCS vapour-prepared silicon wafer surfaces are shown in Figure 7.4.[36] In contrast to the smooth surface features obtained in air for the same kind of substrate surface, clear domains cover quite randomly the whole surface of observation. If, for now, we assume that these domains are bubbles, the base radius and height of these bubbles are in the range between 50 and 400 nm and 20 and 80 nm, respectively. The size range of these bubbles and the manner of

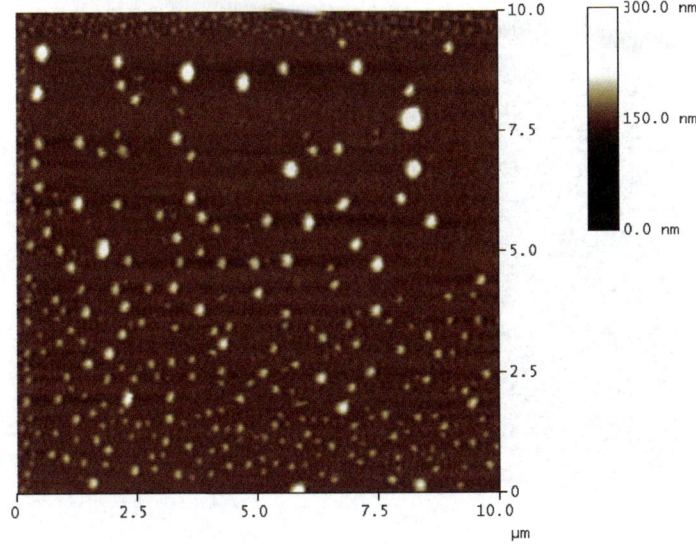

Figure 7.4   Left: TMAFM image of silicon surface modified with TMCS vapour for 20 min. Right: height variation of cross-section in saturated CO_2 solution: $10 \, \mu m \times 10 \, \mu m$, height = 300 nm. (Reproduced with permission from the American Chemical Society, *Journal of Physical Chemistry B*, 2003, **107**, 25, Figure 7.)

their distribution are similar to those observed by other researchers. However, the bubbles shown in the image in Figure 7.4 are much more densely populated and much more evenly distributed over the whole surface of observation compared with those observed by Ishida *et al.*[34]

TMCS/cyclohexane solution-prepared silicon wafer surface. The TMAFM image for the "rough" surface obtained by TMCS/cyclohexane solution methylation is shown in Figure 7.5. In comparison with the bubbles for the smooth surface, one major difference is that bubbles on the rough surface are much less densely populated and the bubble sizes are larger than those for the smooth surface. The bubble base radius and height are in the range of 150–400 nm and 60–200 nm, respectively.

Note that a distinct surface pattern is reflected in the baselines in the cross-section profile, evidenced by small individual peaks. These different features of bubble size and distribution reflect the effects of surface roughness on bubble formation. Note that the same degree of phase shift was observed over the domains as in the case of the TMCS vapour-prepared substrate.

7.2.4 Images of Bubble Coalescence

Consecutive imaging using the TMAFM technique at time intervals over a long time scale was performed to provide further evidence as to the character of the

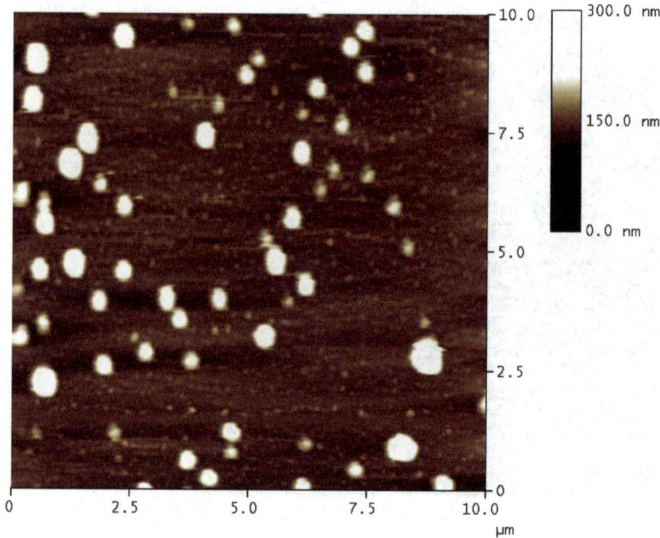

Figure 7.5   Left: TMAFM image of silicon surface modified with TMCS solution for 20 min. Right: height variation of cross-section in saturated CO_2 solution: $10\,\mu m \times 10\,\mu m$, height $= 300\,nm$. (Reproduced with permission from the American Chemical Society, *Journal of Physical Chemistry B*, 2003, **107**, 25, Figure 9.)

domains observed by the TMAFM imaging. The substrates used were the TMCS/cyclohexane solution-prepared silicon wafer ($\theta_w = 88°$, RMS roughness $= 2.7\,nm$). The early stage of the TMAFM imaging process was the same as for normal imaging. However, after the completion of the first TMAFM image, the fluid cell was flushed gently with Milli-Q water every 20 min using a $10\,cm^3$ syringe, and the images were followed by TMAFM scanning. This whole process lasted for more than 20 h. The first image was similar to the bubble image shown in Figure 7.5; five sets of bubble images obtained thereafter are presented in Figure 7.6.

First, one can see in this series of images that, after long-time TMAFM scanning, the bubbles that initially distribute evenly over the whole surface "line up" around a vertical line in the observation area. This phenomenon seems to be evident in TMAFM images reported elsewhere, albeit indistinctly. It can be reasonably assumed that such a pattern of behaviour may be attributed to some fluid–bubble perturbation by the cantilever during the extended horizontal movement or scanning in the fluid cell. This perturbation may also contribute to bubble coalescence.

However, the single most important observation from these images is the process of bubble coalescence and the disappearance of an individual bubble from the solid–solution interface surface, behaviour that is characteristic of gas bubbles in aqueous solutions. For example, the smaller bubble (in the broken-line circle) becomes even smaller from image (a) to image (d) in Figure 7.6, and

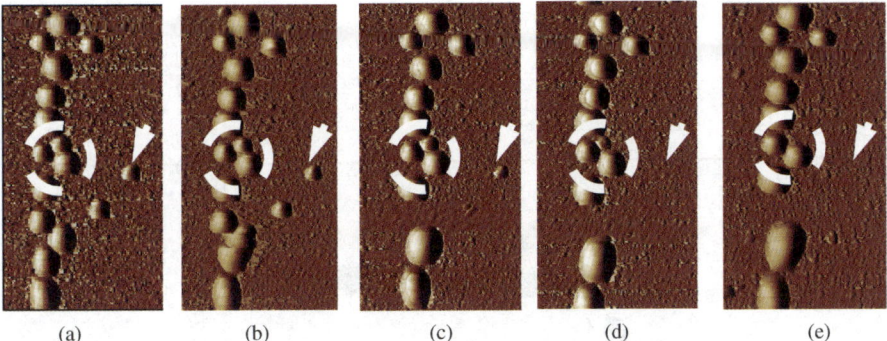

(a)	(b)	(c)	(d)	(e)

Figure 7.6 TMAFM bubble coalescence images: (a) made at 20 h after CO_2-saturated water was injected into the fluid cell. (b)–(e) made at time intervals of 20 min after (a). Scan area: $10\,\mu m \times 5\,\mu m$, height $= 300\,nm$. (Reproduced with permission from the American Chemical Society, *Journal of Physical Chemistry B*, 2003, **107**, 25, Figure 10.)

finally "dissolves" or merges with the neighbouring bubble so that the latter becomes a larger bubble. An individual small bubble (identified by the arrow) becomes increasingly smaller from image (a) to image (c) in Figure 7.6 and finally disappears from image (d). The observed phenomenon, namely that a smaller bubble becomes smaller or dissolves while the neighbouring bigger bubble becomes larger, is quite similar to "Ostwald ripening". It is the first time that such a very small bubble ripening or coalescence process has been observed by TMAFM imaging.

7.2.5 Nature of Domains

Regarding the nature of the domains, several pieces of evidence identify them as gas bubbles. (1) The first important piece of evidence is the large phase shift (approximately 50°) observed over the domains during phase imaging (see Figure 7.7). This large phase shift indicates that the domains are soft and different in nature from the hydrophobised solid surfaces. A typical view of the phase shift across a bubble, generated from TMAFM data files, is shown in Figure 7.7. From this figure, one can see that, although the cross-section of the bubble can be fitted to a spherical cap curve (note the different scales for the base and the height), a constant phase value of around 50° across virtually the whole distance of the selected bubble was seen. (2) Because TMCS, a small molecule (MW = 108.64), is attached through –Si–O–Si– bonds with the silicon wafer surface, it is very unlikely that these adsorbed TMCS molecules after vapour treatment of the solid surface will accumulate into large domains on the substrate surface when it is immersed in the fluid cell. (3) The evidence for bubble coalescence or Ostwald ripening (see Figure 7.6) also supports very strongly that the domains are gas bubbles. These domains would only coalesce

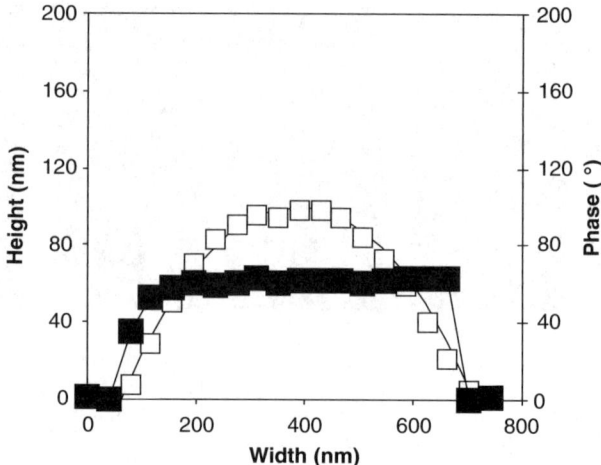

Figure 7.7   Cross-section of a bubble height and phase image profile created from TMAFM image file (TMCS vapour treatment silicon wafer, supersaturated CO_2 solution). Open squares, height image; filled squares, phase image. (Reproduced with permission from the American Chemical Society, *Journal of Physical Chemistry B*, 2003, **107**, 25, Figure 11.)

or dissolve if they are indeed gas bubbles; domains of polymeric material would not be expected to coalesce nor change in size over the time frame of observation. Furthermore, we note that in very strongly degassed solution the structures observed did not change with time. Their mechanical properties are different from bubbles. The force *versus* distance behaviour is indicative of the presence of bubbles on the solid surface.

The force measurements for a colloid probe approaching and then separating from the bubble-covered, solution-methylated silicon wafer surface were obtained by using a methylated silica colloid probe (radius = 8 μm; Figure 7.8).

We have described the procedures elsewhere.[54] The long-range jump distance (>70 nm) for the approaching probe, coupled with the high adhesion force during retreat, is characteristic of the presence of gas bubbles on these surfaces. The stepwise character of the force *versus* distance curve is consistent with bubble bridging for this rough surface.

7.2.6   Bubble Formation on Solid–Water Interfaces

TMAFM images show that there is no nanobubble formation on smooth *hydrophilic* surfaces. This is in agreement with earlier observations in the literature but in contrast to a recent report that nanobubbles are present on freshly cleaved (hydrophilic) mica surfaces.

For dehydroxylated surfaces, there was no nanobubble formation. Bubble formation does not occur on this smooth surface with intermediate hydrophobicity ($\theta_w = 42°$) and a small degree of hysteresis.

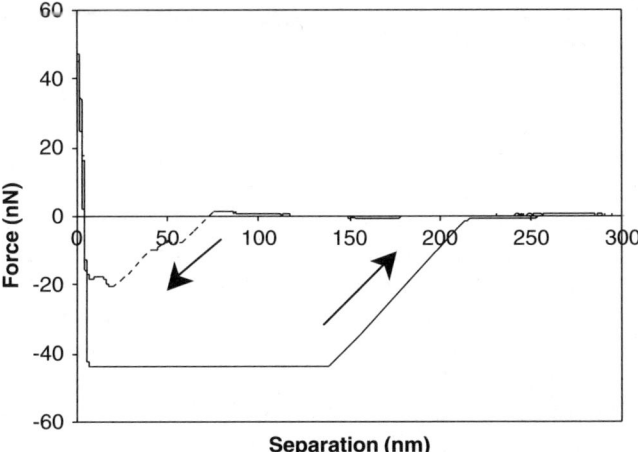

Figure 7.8   Force–separation curves measured upon a hydrophobised silica colloid probe approaching (dashed line) and retreating from (solid line) a small bubble formed on a methylated silicon wafer surface. (Reproduced with permission from the American Chemical Society, *Journal of Physical Chemistry B*, 2003, **107**, 25, Figure 12.)

However, when the surface hydrophobicity increased to the level of the TMCS vapour methylated silicon wafer substrate ($\theta_{wadv} = 74°$), with approximately the same physical roughness and degree of hysteresis as that of dehydroxylated surfaces, bubble formation occurs. This evidence is important for two reasons. First, it demonstrates that the degree of hydrophobicity is important for bubble formation. This agrees with early theoretical analyses, which showed that quite apart from the geometry and size of surface sites, hydrophobicity is essential for bubble formation on solid surfaces. Second, considering that the physical roughness or "defects" are also only of nanoscale size, the observation of bubble formation on such smooth hydrophobic surfaces suggests that small defects in the nanometre range are large enough to act as physical sites for bubble growth on solid surfaces. There was no detectable bubble formation in the bulk solution and thus sites on the solid surface favour bubble formation. The small degree of hysteresis evident for these vapour-treated silicon wafers is sufficient to pin the three-phase contact line at the surface heterogeneity.

There are distinct differences in bubble size and distribution observed for bubble images obtained for the TMCS vapour treated surface and the TMCS in cyclohexane solution prepared surface. The differences in bubble formation are due to the differences in surface roughness and hysteresis of the two kinds of substrates. The TMCS/cyclohexane-treated surface has a much higher surface roughness and larger hysteresis compared with the TMCS vapour-treated surfaces. We speculate that there may well be a larger number of surface sites on the former surface that stimulate bubble formation, perhaps leading to bubble growth and coalescence between adjacent, closely spaced sites. At this point we cannot identify these sites or their distribution with any more certainty.

7.2.7 Contact Angles and Line Tension

The water contact angles (or air contact angles) of the small bubbles on the methylated substrate surfaces are much greater (smaller) than the contact angles measured by the sessile drop methods. The contact angles of the small bubbles can be determined from the cross-sectional profile of each bubble. This may be extracted from the data file of a bubble examined by TMAFM. For Figure 7.7, the advancing air contact angles of bubbles were found statistically to lie in the range 23–30°. Similar observations were made by Ishida *et al.*[34] in their study. The radius of curvature of the AFM tips used falls between 5 and 20 nm. Following detailed work reported elsewhere on fluid droplets, our very small bubble images are not distorted under our experimental conditions.

The difference between these macroscopic and microscopic contact angles may be linked to the influence of line tension. The Young angle is modified by a line tension term when the droplet or bubble size is small:

$$\cos\theta = \cos\theta_Y - \frac{\tau}{\gamma_{lv} R} \tag{7.28}$$

where θ is the actual contact angle that the very small bubble forms with the substrate; θ_Y is the Young contact angle; γ_{lv} is the liquid–vapour surface tension; $1/R$ is the local curvature of the bubble base on solid surface; and τ is the line tension. It should be pointed out that eqn (7.28) reflects only line tension effects, but does not consider the surface roughness and texture effects on the actual macroscopic or microscopic contact angle.[55]

Using eqn (7.28), and examining some tens of bubbles, the dependence of local, microscopic advancing air contact angles on the corresponding curvatures of the very small bubbles (examined from above) is shown in Figure 7.9.

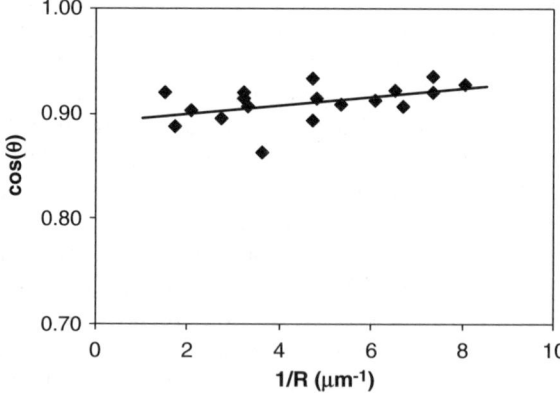

Figure 7.9 Cosine of contact angles of nanobubbles plotted against the corresponding local curvatures based on the TMAFM nanobubble images. The value of line tension calculated from the slope is approximately -2.9×10^{-10} N. (Reproduced with permission from the American Chemical Society, *Journal of Physical Chemistry B*, 2003, **107**, 25, Figure 13.)

The line tension of the bubbles, calculated from the gradient of the regression line, is -3×10^{-10} N. This value is similar in magnitude to the reported values of line tension for small droplets,[56,57] very close to the values determined by Scheludko *et al.*[58] and within the range anticipated theoretically.[59] The negative line tension of the three-phase contact line acts to flatten the bubbles, reducing the Laplace pressure and thus stabilising the very small bubbles formed at these solid–water interfaces. We note that the large bubble contact angle is not re-covered as R goes to infinity in Figure 7.9, behaviour noticed elsewhere.[56,57] We cannot be certain as to the reason at this juncture, but suspect that it may reflect differences in local surface texture for small bubbles compared with large ones with, perhaps, additional mechanisms contributing as well.

7.2.8 Kinetics of CO_2 Gas Adsorption Using a QCM

QCM data provide very valuable information on the kinetic processes involved in the early stages of bubble formation and growth on a hydrophobic surface, before the gas bubbles can be imaged by AFM. Two stages are observed in the frequency shift *versus* time curve in Figure 7.10, following an induction time of approximately 50 s: a first stage of slow bubble growth taking place up to 150 s followed by a fast second stage. This CO_2 gas adsorption process involves the diffusion of gas molecules from the bulk liquid to the interface where they subsequently adsorb at surface sites, displacing water molecules in the process:

$$CO_2 \text{ (bulk)} + H_2O \text{ (surface)} \rightarrow CO_2 \text{ (surface)} + H_2O \text{ (bulk)}$$

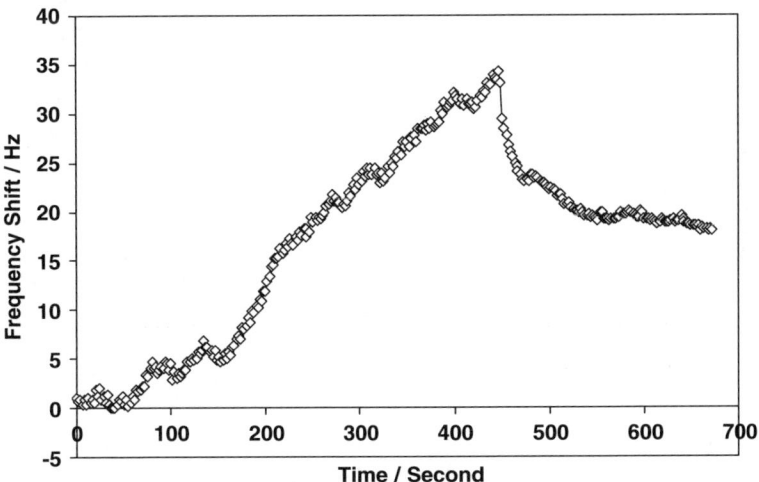

Figure 7.10   Typical time course of gas adsorption/bubble formation at a water–solid interface monitored by a QCM. (Reproduced with permission from the Royal Society of Chemistry from ref. 37.)

We have noted that for gas adsorption to occur, chemical or physical heterogeneities must exist on a hydrophobic surface, *i.e.* surface sites or Harvey nuclei. It is known that heterogeneous nucleation occurs more readily than homogeneous nucleation.[38,60–62] For a hydrophobised solid surface immersed in a gas-saturated water solution, gas molecules from the bulk water will preferentially diffuse from the bulk to the interface and thence to the pre-existing gas cavities, rather than onto smooth hydrophobic surfaces, due to the "catalytic role" of pre-existing gas cavities in bubble nucleation.[60,63] This is the initial or slow stage of gas adsorption.

The second or fast stage of gas adsorption in Figures 7.10 and 7.11 may be associated with the diffusion of gas molecules into gas bubbles formed on surface sites.[37] As the diffusion of CO_2 molecules into the small gas bubbles proceeds, gas bubbles are not confined any more to the surface site and are now free to grow. Bubble coalescence may even occur among neighbouring bubbles on the solid surface. Eventually, as the QCM data show, they detach from the surface.

We note in Figure 7.11 that there is a clear two-stage process with a definite logarithmic dependence on time. We propose that for a CO_2 molecule to initially move from a position in the interface to a surface site the jump

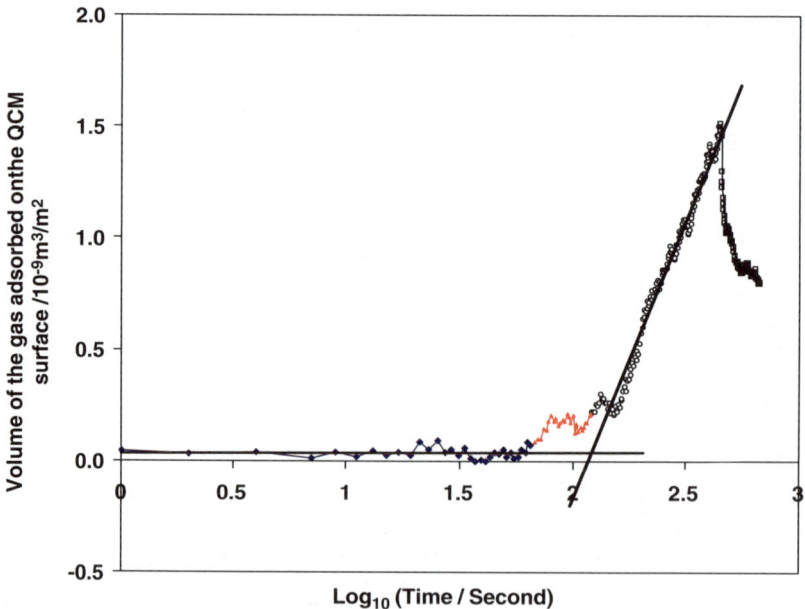

Figure 7.11   Volume of CO_2 gas adsorbed per unit solid surface as a function of time. (Reproduced with permission from the Royal Society of Chemistry from ref. 37.)

frequency[64] is given by

$$\text{Jump frequency} = v\left(\frac{-\Delta E}{kT}\right) \tag{7.29}$$

where v is the thermal vibration of the molecule and ΔE is the activation energy of the process.[65-67] This process is directly analogous to the adsorption/desorption model of molecular displacement occurring within the three-phase zone during dynamic wetting.[68,69]

By analogy with the growth of oxide films,[65] and activation of semi-conductors,[70] the rate of adsorption of CO_2 is given by

$$\frac{d\Gamma}{dt} = Av\left(\frac{-\Delta E_1}{kT}\right) \tag{7.30}$$

where A is a constant.

Many gases exhibit an activation energy which increases with surface coverage according to[64]

$$\Delta E_1 = b_1 N_A \Gamma$$

where b_1 is a constant, Γ is the adsorption density in mol m^{-2} and N_A is Avogadro's number. Thus

$$\frac{d\Gamma}{dt} = Av \exp\left(\frac{-b_1 N_A \Gamma}{kT}\right) \tag{7.31}$$

Integration over the limits ($t = 0$, $\Gamma = 0$) and ($t = t$, $\Gamma = \Gamma$) yields

$$\Gamma = \frac{kT}{b_1 N_A} \ln\left(\frac{b_1 N_A Avt}{kT}\right) + 1 \tag{7.32}$$

If $t \gg kT/b_1 N_A Av$, as suggested in the literature,[58] then

$$\Gamma = \frac{kT}{b_1 N_A} \ln t + \frac{kT}{b_1 N_A} \ln \frac{b_1 N_A Av}{kT} \tag{7.33}$$

or

$$\Gamma = k_1 \ln t + \Gamma_1$$

where k_1 and Γ are constants at constant temperature and are defined through eqn (7.32).

We observe that CO_2 adsorption follows a logarithmic dependence on time up to a rather low surface coverage, with a small rate constant and a large activation energy. Then there is a transition where the activation energy decreases sharply to ΔE_2 with a much increased rate constant. We suggest that this corresponds to a critical bubble size/aggregation of CO_2 molecules at a surface site, where further adsorption occurs simply by CO_2 molecules crossing the liquid–gas interface from their interfacial position, having diffused from the bulk. This second step is expected to have a much lower activation energy than the first. Thus

$$\Delta E_2 = b_2 N_A \Gamma \quad \text{and} \quad b_2 < b_1 \quad \text{with} \quad \Gamma = k_2 \ln t + \Gamma_2$$

This two-step process is observed experimentally (Figure 7.11).

7.3 Concluding Remarks

The rate of adsorption of CO_2 molecules dissolved in aqueous solution onto a hydrophobised silica surface was investigated using a QCM. The results of this investigation were compared with those obtained from TMAFM under the same experimental conditions. The QCM results represent the early stage of CO_2 gas adsorption (<20 min) before CO_2 gas bubbles adsorbed on the surface can be directly observed by TMAFM. The QCM results confirmed our observation from TMAFM imaging that CO_2 gas molecules present in solution only adsorb on silica when its surface is hydrophobic and heterogeneous. More importantly, the results showed that gas adsorption/bubble growth undergoes two consecutive kinetic processes: a slow and a fast adsorption process. The slow process may be associated with the diffusion of gas molecules from an interfacial region to surface sites (Harvey nuclei). Once a critical bubble/aggregate size is reached, further gas adsorption takes place.

The debate concerning the origin of large attractive forces, observed between hydrophobic surfaces, has continued for many years. In our experiments, we have found that in the case of very smooth, uniform hydrophobic (dehydroxylated) and hydrophilic surfaces, classical DLVO theory can explain the interactions between these surfaces. Surface heterogeneity, both chemical and physical, and hydrophobicity promote the formation of gas bubbles. These very small surface bubbles are predicted to change both the van der Waals and electrostatic forces acting between the particles. The methylated silica surfaces we have used in this study are hydrophobic and display both chemical and physical heterogeneity. The data presented here show a clear link between the level of dissolved gas present in solution and the related colloid stability of the particles. In contrast, the colloidal stability of dehydroxylated and hydrophilic silica particles is unaffected by dissolved gas levels. The colloidal interactions have been described using an appropriate modification of DLVO theory. The analysis shows that surface-to-surface interactions between particles are dominated by the presence of very small, protruding surface bubbles.

Acknowledgement

Financial support from the Australian Research Council Special Research Centre Scheme is gratefully acknowledged.

References

1. B. Vincent, *J. Colloid Interface Sci.*, 1973, **42**, 270.
2. G. Fleer, M. Cohen Stuart, J.M.H.M. Scheutjens, T. Cosgrove and B. Vincent, *Polymers at Interfaces*, Chapman and Hall, London, 1993; L. Ter Minassian-Saraga, M. Adler, A. Barraud, N. V. Churaev, D. F. Eaton, H. Kuhn, M. Misono, D. Platikanov, J. Ralston, A. Silberberg, B. Vincent and J. N. Zemel, *Pure Appl. Chem.*, 1994, **66**, 1667.
3. R. Boyle, *Philos. Trans. R. Soc. London*, 1670, **5**, 2039.
4. E.N. Harvey, D.K. Barnes, W.D. McElroy, A.H. Whiteley, D.C. Pease and K. W. Cooper, *J. Cell. Comp. Physiol.*, 1944, **24**, 1–22; E. N. Harvey, D. K. Barnes, W. D. McElroy, A. H. Whiteley, D. C. Pease and K. W. Cooper, *J. Cell. Comp. Physiol.*, 1944, **24**, 23–24.
5. E.N. Harvey, *Bull. N. Y. Acad. Med.*, 1945, 505.
6. E.N. Harvey, K.W. Cooper and A.H. Whiteley, *J. Am. Chem. Soc.*, 1946, **68**, 2119.
7. E.N. Harvey, W.D. McElroy and A.H. Whiteley, *J. Appl. Phys.*, 1947, **18**, 162.
8. N.F. Bunkin, O.A. Kiseleva, A.V. Lobeyev, T.G. Movchan, B.W. Ninham and O.I. Vinogradova, *Langmuir*, 1997, **13**, 3024.
9. S. Wrobel, *Mine Quarry Eng.*, 1952, 313.
10. S.D. Lubetkin, *Langmuir*, 2003, **19**, 2575.
11. M. R. Urban, PhD thesis, University of London, 1978.
12. S.D. Lubetkin, *J. Chem. Soc. Faraday Trans.*, 1989, **1**, 1753.
13. R.F. Considine, R.A. Hayes and R.G. Horn, *Langmuir*, 1999, **15**, 1657.
14. L. Meagher and V.S.J. Craig, *Langmuir*, 1994, **10**, 2736.
15. Z.A. Zhou, X. Zhenghe and J.A. Finch, *J. Colloid Interface Sci.*, 1996, **179**, 311.
16. J. Mahnke, J. Stearnes, A. Hayes, D. Fornasiero and J. Ralston, *Phys. Chem. Chem. Phys.*, 1999, **1**, 2793.
17. J.L. Parker, P.M. Claesson and P. Attard, *J. Phys. Chem.*, 1994, **98**, 8468.
18. G.E. Yakubov, H.-J. Butt and O.I. Vinogradova, *J. Phys. Chem. B*, 2000, **104**, 3407.
19. E.N. Harvey, D.K. Barnes, W.D. McElroy, A.H. Whiteley, D.C. Pease and K.W. Cooper, *J. Cell. Comp. Physiol.*, 1944, **24**, 1.
20. E.N. Harvey, A.H. Whiteley, W.D. McElroy, D.C. Pease and D.K. Barnes, *J. Cell. Comp. Physiol.*, 1944, **24**, 23.
21. E.N. Harvey, K.W. Cooper and A.H. Whiteley, *J. Am. Chem. Soc.*, 1946, **68**, 2119.
22. E.N. Harvey, W.D. McElroy and A.H. Whiteley, *J. Appl. Phys.*, 1947, **18**, 162.
23. N. Ishida, T. Inoue, M. Miyahara and K. Higashitani, *Langmuir*, 2000, **16**, 6377.
24. S.T. Lou, Z.Q. Ouyang, Y. Zhang, X. Li, J. Hu, M.Q. Li and F.J. Yang, *J. Vac. Sci. Technol. B*, 2000, **18**, 2573.

25. J.W.G. Tyrrell and P. Attard, *Langmuir*, 2002, **16**, 6377.
26. J. Ralston, D. Fornasiero and N. Mishchuk, *Colloids Surf. A*, 2001, **192**, 39.
27. D.R.E. Snoswell, J. Duan, D. Fornasiero and J. Ralston, *J Phys. Chem. B*, 2003, **107**, 2986.
28. M.E. Karaman, D.A. Antelmi and R.M. Pashley, *Colloids Surf. A*, 2001, **182**, 285.
29. H.K. Christenson and P.M. Claesson, *Adv. Colloid Interface Sci.*, 2001, **91**, 391.
30. R.J. Hunter, *Foundations of Colloid Science*, Clarendon Press, Oxford, 1987, ch. 7.
31. S.H. Behrens, D.I. Christl, R. Emmerzael, P. Schurtenberger and M. Borkovec, *Langmuir*, 2000, **16**, 2566.
32. R.K. Iler, *The Chemistry of Silica: Solubility, Polymerization, Colloid and Surface Properties, and Biochemistry*, Wiley, New York, 1979.
33. S.-T. Lou, Z.-Q. Ouyang, Y. Zhang, X.-J. Li, J. Hu, M.-Q. Li and F.-J. Yang, *J Vac. Sci. Technol. B*, 2000, **18**, 2573.
34. N. Ishida, T. Inoue, M. Miyahara and K. Higashitani, *Langmuir*, 2000, **16**, 6377.
35. (a) J.W.G. Tyrrell and P. Attard, *Phys. Rev. Lett.*, 2001, **87**, 176104/1; (b) J.W.G. Tyrrell and P. Attard, *Langmuir*, 2002, **18**, 160.
36. J. Yang, J. Duan, D. Fornasiero and J. Ralston, *J. Phys. Chem. B*, 2003, **107**, 6139.
37. J. Yang, J. Duan, D. Fornasiero and J. Ralston, *Phys. Chem. Chem. Phys.*, 2007, **42**, 6327.
38. P.M. Wilt, *J. Colloid Interface Sci.*, 1986, **112**, 530.
39. S.D. Lubetkin and M.R. Blackwell, *J. Colloid Interface Sci.*, 1988, **26**, 610.
40. M.W. Carr, A.R. Hillman, S.D. Lubetkin and M.J.J. Swann, *J. Electroanal. Chem. Interfacial Electrochem.*, 1989, **267**, 313.
41. M.W. Carr, A.R. Hillman and S.D. Lubetkin, *J. Colloid Interface Sci.*, 1995, **169**, 135.
42. D.A. Buttry and M.D. Ward, *Chem. Rev.*, 1992, **92**, 1355.
43. T. Sato, T. Serizawa, F. Ohtake, M. Nakamura, T. Terabayashi, Y. Kawanishi and Y. Okahata, *Biochim. Biophys. Acta*, 1998, **1380**, 82.
44. Y. Ebara, K. Itakura and Y. Okahata, *Langmuir*, 1996, **12**, 5165.
45. W. Gong, J. Stearnes, D. Fornasiero and R.A. Hayes, *Phys. Chem. Chem. Phys.*, 1999, **1**, 2799.
46. F. Grieser, R.N. Lamb, G.R. Wiese, D.E. Yates, R. Cooper and T.W. Healy, *Radiat. Phys. Chem.*, 1984, **23**, 43.
47. B.V. Derjaguin, N.V. Churaev and V.M. Muller, in *Surface Forces*, ed. J. A. Kitchener, Plenum Press, New York, 1987.
48. D.N.L. McGown and G.D. Parfitt, *J. Phys. Chem.*, 1967, **71**, 449.
49. E.P. Honig, G.J. Roebersen and P.H. Wiersema, *J. Colloid Interface Sci.*, 1971, **36**, 97.
50. L.A. Spielman, *J. Colloid Interface Sci.*, 1970, **33**, 562.
51. A.M. Puertas and F.J. Nieves, *J. Colloid Interface Sci.*, 1999, **216**, 221.
52. S. Usui and E. Barouch, *J. Colloid Interface Sci.*, 1989, **137**, 281.

53. M. Zembala and Z. Adamczyk, *Langmuir*, 2000, **16**, 1593.
54. J. Ralston, I. Larson, M. Rutland, A. Feiler and M. Kleijn, *Pure Appl. Chem.*, 2005, **77**, 2149.
55. J. Drelich, J.D. Miller and R.J. Good, *J. Colloid Interface Sci.*, 1996, **179**, 37.
56. E.A. Hemmingsen, *J. Appl. Phys.*, 1975, **46**, 213.
57. T. Pompe and S. Herminghaus, *Phys. Rev. Lett.*, 2000, **85**, 1930.
58. A. Scheludko, B. Toshev and B. Bogadiev, *J. Chem. Soc., Faraday Trans.*, 1976, **72**, 2815.
59. F.P. Buff and H.T. Saltsburg, *J. Phys. Chem.*, 1957, **26**, 23.
60. S.F. Jones, G.M. Evans and K.P. Galvin, *Adv. Colloid Interface Sci.*, 1999, **80**, 27.
61. R. Cole, in *Advances in Heat Transfer*, ed. T. F. Irvine Jr and J. P. Hartnett, Academic Press, New York, 1974, Vol. **10**, p. 85.
62. M. Blander and J.L. Katz, *AIChE J.*, 1975, **21**, 833.
63. P.W. Atkins, *Physical Chemistry*, Oxford University Press, Oxford, 1994.
64. G.A. Somorjai, *Principles of Surface Chemistry*, Prentice-Hall, Englewood Cliffs, NJ, 1972.
65. N.F. Mott, *Trans. Faraday Soc.*, 1940, **36**, 472.
66. E.A. Moelwyn-Hughes, *in Physical Chemistry*, Pergamon, Oxford, 2nd edn, 1961, ch. 1.
67. A.W. Adamson, *Physical Chemistry of Surfaces*, Wiley Interscience, New York, 5th edn, 1990.
68. T.D. Blake, in *Wettability*, ed. J. C. Berg, Marcel Dekker, New York, 1993, ch. 5.
69. J.G. Petrov, M. Schneemilch and J. Ralston, *Langmuir*, 2003, **19**, 2795.
70. J. Ralston and T.W. Healy, *Int. J. Min. Proc.*, 1980, **7**, 175.

Chapter 8

Heteroflocculation Studies of Colloidal Poly(*N*-isopropylacrylamide) Microgels with Polystyrene Latex Particles: Effect of Particle Size, Temperature and Surface Charge

Martin J. Snowden,* Louise H. Gracia and Hani Nur

MEDWAY SCIENCES, UNIVERSITY OF GREENWICH, CENTRAL AVENUE, CHATHAM MARITIME, KENT ME4 4TB, UK

Abstract

The heteroaggregation behaviour of thermosensitive cationic poly(*N*-isopropyl-acrylamide) [poly(NIPAM)] microgels and anionic polystyrene (PS) latex particles has been investigated with respect to particle size, surface charge, deformability of the particle, temperature and concentration. Microgels and PS latex particles have been characterised. Turbidimetric analysis and scanning electron microscopy have been employed to determine the state of aggregation of mixtures of the two types of particles.

PS latex particles of varying size (100–800 nm) have been studied in mixtures with two cationic microgels. Results show that the ionic interaction between oppositely charged deformable and non-deformable particles can be manipulated by changing the particle size, concentration and temperature of the particles in the mixture. It has been found that aggregation is induced in the mixtures containing the largest of the PS latex particles by the lowest concentration of microgel. It is also reported that the larger the latex the less is the influence of temperature on the stability of the mixed particulate system.

* Corresponding author.

New Frontiers in Colloid Science: A Celebration of the Career of Brian Vincent
Edited by Simon Biggs, Terence Cosgrove and Peter Dowding
© The Royal Society of Chemistry 2008

The heteroflocculation/aggregation behaviour is influenced by the relative concentration of the two oppositely charged species. The change from a dispersed to an aggregated system is most likely continuous and the relative concentration at which aggregation begins to occur is somewhere between a 1 : 1 ratio and 10 : 1 excess of the cationic microgel. This concentration effect can be used to initiate aggregation at temperatures below the volume phase transition temperature of the poly(NIPAM) microgels.

8.1 Introduction and Background

Professor Brian Vincent's successful career has spanned several decades and has covered numerous aspects of the study of polymer and colloid science. One particular area of interest in this field, to which he has contributed enormously, is concerned with the stability of colloidal dispersions. Brian Vincent's research has covered many aspects of colloid stability, incorporating the stability of emulsions[1] and suspensions,[2] the theory behind and the application of polymeric[3–5] and electrostatic[6] stabilisation, and also depletion[7,8] temperature[9] or electrolyte[10] induced flocculation/aggregation, covering both homofloccula-tion[10] and heteroflocculation.[11,12] His research has also covered the topic of colloidal microgel particles,[9,13,14] investigating their fundamental physico-chemical characteristics, including microgel stability, through to their current and potential applications. In the study of colloidal dispersions, many different physicochemical factors have to be investigated in an attempt to understand their properties and to optimise their utility in industrial processes.[15] One important area is the study of the stability of particulate matter when dispersed within a solvent environment. The factors which contribute most to the overall nature of a colloidal system are: particle size, shape and flexibility; the nature of the particle surface; and the particle–particle and particle–solvent interactions.[16] Stability is dependent on the interactions that occur between the particles and also between the particles and the solvent. When a system contains only one type of particle, the kind of interactions that predominate will differ from the situation where more than one type of particle coexists within the same isotropic medium.[15] As a result of particle collisions in a dispersion, particles may rebound off one another or aggregate. Aggregation may result in permanent contact which is termed coagulation, or temporary contact which is known as flocculation. When aggregation takes place an unstable colloidal system is the result.

Colloidal microgel particles are a class of cross-linked polymeric particles, often given the prefix of "smart" or "intelligent" materials. These particles have been described as "smart" due to their ability to undergo conformational changes in response to a change in their environmental conditions. Microgel particles are capable of undergoing a reversible conformational transition (volume phase transition, VPT) which brings about rapid changes in their physical properties including large changes in particle size, volume and surface charge density. Microgels are of significant interest from both an academic and industrial perspective with a view to their use in a number of potential applications, for example drug delivery vehicles,[17–19] biosensors[20,21] and in other areas of biotechnology.[22] For poly(*N*-isopropylacrylamide) [poly(NIPAM)] microgels, as charged groups are introduced by initiators onto the particle surface, the dispersion remains stable even when the solvent quality is poor, *e.g.* at high temperature. Applications where a stable colloid can be destabilised by heteroaggregation[23] are of particular interest, *e.g.* for the exploitation of the thermosensitivity of poly(NIPAM) microgels in processes designed to enhance the efficiency of crude oil recovery.[24] The process relies on the charged microgel particles forming stable dispersions below their volume phase transition temperature (VPTT) and flocculating when injected into the ground. Flocculation occurs in response to the higher temperatures prevalent in oil-bearing rock (above the VPTT) and the presence of electrolytes in the pumping material, *e.g.* seawater. The colloidal aggregates block off channels of high permeability enabling the oil in the less permeable areas to be mobilised. The process is reliant on the strength of the aggregation, *i.e.* its ability to endure large shear forces resulting from high pressures. Mixed charge colloidal dispersions may exhibit stronger aggregation characteristics than simple van der Waals forces when electrostatic attractive forces also contribute to the robustness of the resulting aggregates.

The colloid stability of poly(NIPAM) microgels has been investigated as a function of electrolyte and free polymer concentration as well as temperature.[9,13] The dispersions were found to remain colloidally stable in electrolyte concentrations up to about $1.0 \, mol \, dm^{-3}$ NaCl, at room temperature. This behaviour is attributed to the microgels being swollen with solvent, at room temperature. The van der Waals attraction is eliminated for swollen particles because the particles are predominantly water and therefore possess a Hamaker constant that is matched to the surrounding fluid; hence the driving force favouring aggregation is minimal. In addition, microgels below their VPTT exhibit an element of steric repulsion arising from their "hairy" surface. NIPAM-based microgels have been found to aggregate above their VPTT when sufficient electrolyte was added.[25] The particles in the de-swollen state are electrostatically stabilised against aggregation, but when a microgel dispersion is heated above the VPTT, the particles collapse and a large proportion of the solvent contained within their structure is excluded, resulting in an increase in the particle Hamaker constant. This in turn increases the van der Waals attractive force, and in the presence of relatively high ($>0.05 \, mol \, dm^{-3}$ NaCl) electrolyte concentration, when the length of the electrical double layer (Debye length) surrounding the particles is very short ($<10 \, nm$), aggregation of the particles takes place. Interestingly this aggregation is reversible; when the temperature decreases below the VPTT the particles redisperse. However, mixtures of *oppositely charged*, swollen microgel particles may still undergo heteroaggregation below their VPTT at low electrolyte concentrations, due now to interparticle *electrostatic* attraction.[26] The heteroaggregation characteristics of mixed colloidal dispersions of anionic poly(NIPAM) and cationic poly(NIPAM)-*co*-(4-vinylpyridine) have been examined in relation to their relative concentrations.[25] The degree of aggregation/dispersibility has been analysed in relation to environmental solution conditions of pH and electrolyte concentration using turbidimetric and transmission electron microscopy (TEM) data. It has been shown that heteroaggregation in systems of variable charge and charge density is influenced by a complex balance of forces, both electrostatic and steric, and is affected significantly by the environment of the particles. The data reported in the study also show that without the stabilising effect of the electrolyte screening the charges, aggregation occurs readily at 20 °C except when the cationic microgel was in large excess. Under these conditions the effect of relative concentration promoted the formation of a stable dispersion of hetero-macroparticles at elevated temperatures. It has also been shown that aggregation under various conditions of pH, electrolyte, temperature, charge density and relative concentration can be induced by the careful manipulation of any one or more of the environmental conditions.

When the system undergoing aggregation is composed of different types of particles, the process is known as heteroaggregation, heterocoagulation or heteroflocculation. Heterocoagulation is generally used to describe permanent contact between particles and hence irreversibility. Heteroflocculation and heteroassociation are regarded as defining a reversible or temporary association between particles, and heteroaggregation is a general term applied to particle aggregation. Aggregation of particles with varying composition, charge or size has been shown to be important in many industrial applications, including mineral floatation,[27] cell recovery,[14] stability of emulsions,[28] synthesis of engineering ceramics[29] and waste water treatment.[30] Heteroaggregation, however, is not as extensively used as homoaggregation. This may be mainly due to the relatively complex interactions between dissimilar particles that the classical DLVO theory cannot account for.[31] The particles can be different in variety of ways, for example composition and shape, surface potential and surface charge. Interactions are more likely to occur between particles when the surfaces are dissimilar than the particles of the same type which is the phenomenon of homoaggregation. Several studies have been carried out on systems containing oppositely charged particles.[32] It was shown that the relative size of the particles determines the morphology of aggregation: if there is a significant difference in size, the small particles are adsorbed onto the large particles but if particles have comparable size, growth of large fractal clusters may take place.[32] Heteroaggregates can be stable or unstable depending on whether aggregates remain as regular discrete units or whether they exist in large irregular masses in solution resulting in precipitation. This behaviour is dependent on the relative particle sizes and concentration. When the particles are similar in size, the dissimilar particles adsorb in a random manner, forming irregular clusters of particles.[15] If there is a large difference in particle size, the smaller particles adsorb onto the surface of the larger species. This adsorption can result in unstable heteroaggregates through bridging or, if the whole surface of the large particles is covered, this results in the surface properties of the larger particle being similar to the small particles. If this process is repeated for all the larger particles with the excess concentration of the small particles, then the mixed particle aggregates mimic the properties of the smaller particles; hence, when two like particles collide they will rebound and remain free, resulting in stable heteroaggregates.

When combining microgels with an oppositely charged latex, an ionic interaction ensues causing a heteroflocculation.[15] This interaction can be manipulated by temperature or particle concentration as a trigger to switch on or switch off the heteroflocculation mechanism. However, it becomes extremely difficult to explain the processes occurring in heteroaggregating systems displaying a large relative size asymmetry, charge density and also surface charge potentials as well as being subject to variation under different conditions, relative concentration and/or temperature.

8.1.1 Heteroflocculation Studies

This study is concerned with an examination of the interaction of cationic poly(NIPAM) microgel particles with anionic polystyrene (PS) latex particles of varying size producing data which provide information on the interaction of different types of colloidal particles with respect to the effect of particle size, deformability of the particle, temperature and dispersion concentration. Previous studies by Harley *et al.*[33] and Fernández-Barbero and Vincent[6] have considered the importance of heteroflocculation between small and large particles or soft and hard particles, and several other research groups have studied the causes of aggregation in both non-charged/homocharged[34,35] and heterocharged[36,37] systems by experimental and computer simulation techniques.

Colloidal microgels shrink and swell in response to changes in environmental conditions, an example of which is the collapse of poly(NIPAM) particles in response to increasing temperature. When combining microgels with a latex dispersion made up of oppositely charged particles, an ionic interaction ensues inducing heteroflocculation. This interaction can be manipulated by temperature or particle concentration as a trigger to switch on or switch off the heteroflocculation mechanism.

8.2 Experimental

8.2.1 Microgel Synthesis

The chemicals used in this experiment are as follows: NIPAM 97% (Aldrich, UK; lot no. 06315AC), *N,N'*-methylenebisacrylamide (BA) 99% (Aldrich, UK; lot no. 4C T07719EG), 2,2'-azobis(2-methylpropionamidine) dihydrochloride (Aldrich, UK; lot no. 12012DO-113), allylmethacrylate (AMA) (Fluka, Germany; lot no. 4C T07719EG), salinisation solution II (~2 vol.% dimethylsilane in 1,1,1-trichloroethane) (Fluka, Switzerland, lot no. 015401/1) and aqueous dispersions of polystyrene latex (LB-1, 100 nm, Sigma-Aldrich, USA; lot no. 083K0987; LB-5, 460 nm, Sigma-Aldrich, USA; lot no. 124K1709; LB-8, 800 nm, Sigma-Aldrich, USA; lot no. 034K0658).

All the chemicals were obtained from the suppliers and used without further purification. Distilled water was used for the synthesis and dialysis of the microgels.

Two cationic poly(NIPAM) microgels were prepared using two different cross-linkers, BA and AMA. The microgels are denoted as poly(NIPAM/BA) and poly(NIPAM/AMA) respectively. The microgels were synthesised by surfactant-free emulsion polymerisation (SFEP).[27] The reactions were carried out in a fume cupboard. The cationic initiator 2,2'-azobis(2-methyl-propionamidine) dihydrochloride (0.5 g) was dissolved in 800 ml distilled water in a reaction vessel. The vessel was salinated before use by rinsing the flask with the salinisation solution and then heated in an oven for 2 h at $\sim 70\,^{\circ}\mathrm{C}$. The initiator solution was then heated to $70\,^{\circ}\mathrm{C}$, with constant stirring, under a nitrogen atmosphere. Separately 5.0 g of NIPAM and 0.50 g of either BA or AMA were combined in 200 ml of distilled water. This mixture was then added to the reaction vessel containing the initiator solution at $70\,^{\circ}\mathrm{C}$. The reaction was allowed to proceed at $70\,^{\circ}\mathrm{C}$ for 6 h with continuous stirring. The final product was extensively dialysed for several days against distilled water.

8.2.2 Dry Weight Analysis

Three samples of each of the microgels and PS latex dispersions were placed in glass sample vials, heated in an oven at $80\,^{\circ}\mathrm{C}$ and dried to constant mass. The concentration of the microgel/latex was calculated as a w/w percentage from the weight of the hydrated and dried sample. The mean value of the three samples was calculated to give the mass concentration of the microgel/latex in the dispersion. The dry weight was then used to calculate a percentage mass (wt%) of the microgels and PS latex in solution.

8.2.3 Preparation of Anionic Polystyrene Latex and Cationic Microgel Mixed Dispersions

Four sets of Microgel/latex mixed dispersions were prepared from three different anionic PS latexes of varying size (PS 1 100 nm, PS 5 460 nm and PS 8 800 nm) and the two cationic microgels (poly(NIPAM/BA) 560 nm and poly(NIPAM/AMA) 271 nm). The microgel dispersions (0.01% w/w) were thoroughly mixed with a PS latex dispersion (0.01% w/w) of opposing charge in different ratios, examples of which are displayed in Table 8.1. Samples were prepared for each microgel/latex mixture. The pH was kept constant (5.6) for all the mixed dispersions and the total volume was 10 ml for each sample. The samples remained undisturbed for 24 h after preparation at room temperature after which their physical state was determined visually and the aggregation profile was obtained by turbidimetric analysis. The samples were then heated for a period of 3 h above the VPTT of the microgel ($50\,^{\circ}\mathrm{C}$), after which the

Table 8.1 The composition of 100 nm anionic PS and 271 nm cationic pNIPAM mixtures.

Sample	Mixing ratio PS : pNIPAM	PS (mL)	pNIPAM (mL)	% of pNIPAM
1	10:1	9.09	0.91	9.1
2	7:1	8.75	1.25	12.5
3	5:1	8.33	1.67	16.7
4	4:1	8.0	2.0	20
5	3:1	7.5	2.5	25
6	2:1	6.67	3.33	33.3
7	1.5:1	6.0	4.0	40
8	1:1	5.0	5.0	50
9	1:3	2.5	7.5	75
10	1:5	1.64	8.33	83.3
11	1:10	0.91	9.09	90.9

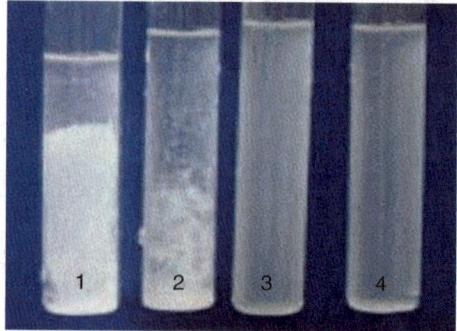

Figure 8.1 Visual observations indicate the stability of the mixed dispersions: 1 and 2 form heterofloccs whereas 3 and 4 are stable.

extent of aggregation was again determined. Figure 8.1 gives a visual presentation of the flocculation experiment.

8.2.4 Dynamic Light Scattering

Microgel/latex dispersions of 0.01% (w/w) were prepared by dilution of the stock dispersions in clean stoppered vials. The hydrodynamic diameter of the particles was measured using a Malvern Zetasizer Nano ZS, equipped with a 4 mW He–Ne laser ($\lambda = 633$ nm) with the detector positioned at 173° to the sample. The size measurement of PS latex particles was performed at room temperature (25 °C). The microgel dispersions were analysed as a function of temperature (10–50–10 °C at 5 °C intervals). The temperature of the dispersions was maintained by a Peltier thermocouple, and the samples were equilibrated at each temperature for 15 min before data collection.

8.2.5 Electrophoretic Mobility and Zeta Potential Measurements

The electrophoretic mobility of the samples was measured using a Malvern Zetasizer Nano ZS. The measurements were made in a background of $1.0 \times 10^{-4}\,\mathrm{mol\,dm^{-3}}$ NaCl solution; the measurements were carried out as a function of temperature (10–50–10 °C and at 5 °C intervals) for both microgels. The microgels were diluted (1 : 4) using the background sodium chloride solution to obtain readings within the operating parameters of the machine and technique. The zeta potential measurements for the latex dispersions were carried out at room temperature (25 °C). Electrophoretic mobility measurements were also made for selected microgel/latex mixtures in an attempt to confirm the presence of the macroparticles observed by scanning electron microscopy.

8.2.6 Turbidimetric Analysis

The mixtures of cationic microgel and anionic latex particles were examined with respect to their flocculation behaviour over a range of relative concentrations and at temperatures below and above the VPTT of the microgel. Each mixture has a dispersibility factor (n) based on the wavelength dependence of the turbidity of the particles: $n = \mathrm{d}\log[\text{turbidity (abs)}]/\mathrm{d}\log[\text{wavelength }(\lambda)]$. The turbidity was measured using a Cary 100 Bio UV-visible spectrophotometer. All mixtures were measured at both 25 and 50 °C (below and above the VPTT of the microgel), at wavelengths of 650, 600, 550, 500 and 450 nm. The more aggregated the mixture the larger the particles/aggregates and hence the lower the turbidity. The lower is the turbidity the less change in absorbance is observed as a function of wavelength, which produces a value of n closer to zero. This allows a range of values of n (0 to -4) to be calculated where values tending towards zero indicate aggregation whilst values tending towards -4 represent a more dispersed system (taking a value of $n \geq -2.00$ as dispersed). The values of n were determined at both 25 °C and 50 °C. The heteroflocculation profile was obtained by plotting the dispersibility values (n) as a function of the relative concentration of poly(NIPAM).

8.2.7 Scanning Electron Microscopy (SEM)

SEM experiments were carried out by Dr Ian Slipper at the School of Science, University of Greenwich at Medway, using a Stereoscan 360 Cambridge SEM instrument. Samples were prepared by drying the sample on a glass slide placed on a carbon slit. The dried samples were then placed in an Edwards Sputter Coater S150B and gold coated for 5 min. The micrographs were obtained by using the SEM operated at 20 kV. The poly(NIPAM) microgel and latex particles were analysed to confirm their monodisperse spherical nature. The mixed dispersions were analysed to show the nature of the dispersion or aggregation under various conditions of concentration and temperature. Samples analysed above the VPTT of the microgel were dried at between 50 and 70 °C on a glass slide.

8.3   Results and Discussion

The microgel dispersions were characterised by dynamic light scattering (DLS) with respect to their size and electrophoretic mobility as a function of temperature. The results are given in Table 8.2; the results obtained from the heating cycle are given as there was no significant hysteresis between this and the cooling cycle.

The poly(NIPAM/BA) showed a size difference of approximately 320 nm between 50 and 10 °C. The poly(NIPAM/AMA) microgel particles have hydrodynamic diameters of 271 and 160 nm at 10 and 50 °C respectively with a size difference of ∼ 111 nm.

The poly(NIPAM/AMA) particles do not achieve the same hydrodynamic size as the poly(NIPAM/BA). This is due partly to the polymerisation process but mainly due to the nature of the AMA cross-linker. The AMA cross-linker contains fewer hydrophilic groups (=CO, –NH) compared to the BA cross-linker, which may decrease the overall microgel–solvent interactions and hence decrease the size of the microgel.

Both the microgels exhibit a large increase in electrophoretic mobility with an increase in temperature (Table 8.2). The charge on the poly(NIPAM) particles is due entirely to the dissociated amidine (H_3N^+) groups localised at the periphery of the particles arising from the cationic initiator. The microgel particles have a much higher mobility in the collapsed state due both to their smaller size creating less resistance, but more importantly to their increased surface charge density brought about by the large decrease in particle volume and therefore surface area.

8.3.1   Scanning Electron Microscopy

Scanning electron micrographs of poly(NIPAM/BA) and poly(NIPAM/AMA) show that both microgels are spherical and monodisperse in nature (Figure 8.2).

Table 8.2   DLS results of pNIPAM (BA) and pNIPAM (AMA) microgels.

Temperature (°C)	pNIPAM (BA)		pNIPAM (AMA)	
	Hydrodynamic diameter (nm)	Electrophoretic mobility ($\times 10^{-8} m^2 s^{-1} V^{-1}$)	Hydrodynamic diameter (nm)	Electrophoretic mobility ($\times 10^{-8} m^2 s^{-1} V^{-1}$)
10	558 ± 5	0.46 ± 0.02	271 ± 3	0.7 ± 0.01
15	558 ± 5	0.49 ± 0.03	275 ± 1	0.79 ± 0.04
20	555 ± 7	0.64 ± 0.03	270 ± 1	0.88 ± 0.10
25	549 ± 2	0.68 ± 0.08	264 ± 5	0.97 ± 0.04
30	539 ± 3	0.89 ± 0.01	248 ± 2	1.13 ± 0.05
35	490 ± 3	1.08 ± 0.05	179 ± 2	2.29 ± 0.03
40	329 ± 2	1.77 ± 0.02	167 ± 1	2.95 ± 0.01
45	278 ± 2	2.36 ± 0.12	166 ± 3	3.61 ± 0.06
50	273 ± 3	2.85 ± 0.16	163 ± 2	3.74 ± 1.71

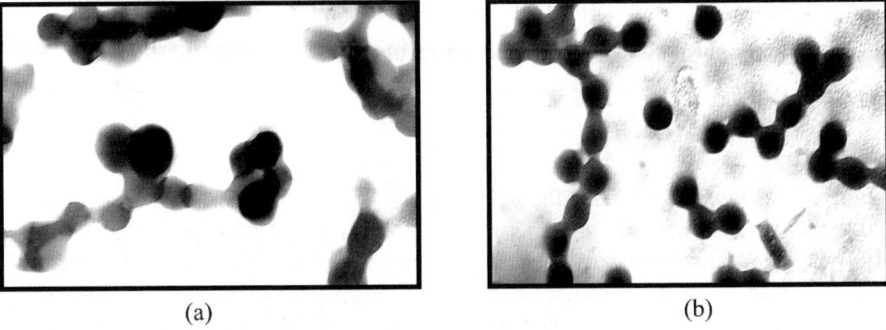

(a)   (b)

Figure 8.2 SEM images displaying cationic poly(NIPAM) microgels: (a) poly (NIPAM/BA); (b) poly(NIPAM/AMA).

Figure 8.3 SEM image of anionic PS 5 latex particles.

Table 8.3 Particle diameter and zeta potential of anionic PS latex particles.

Latex	Particle diameter (nm)	Zeta potential (mV)
PS 1	109 ± 5	-43 ± 3
PS 5	466 ± 3	-44 ± 5
PS 8	804 ± 4	-35 ± 3

8.3.2   Characterisation of Anionic PS Latex Particles

The anionic latex particle were characterised with respect to their particle size and zeta potential using DLS and SEM. The DLS (Table 8.3) and SEM (Figure 8.3) results meet the specification provided by the supplier.

8.3.3   Heteroflocculation of Microgel/Latex Mixtures

In the subsequent sections the heteroflocculation behaviour of each microgel/
latex mixture is outlined individually in relation to their heteroflocculation
profile and SEM results.

8.3.3.1   Cationic Poly(NIPAM) Microgel (271nm) and PS Latex (800nm).
Mixed particle dispersions with low relative microgel to latex concentrations
showed destabilisation and therefore aggregation (n tends to 0; Figure 8.4).
Visual observation of the dispersion at this point provided a clear indication
that aggregation had occurred (Figure 8.5). SEM images show that the

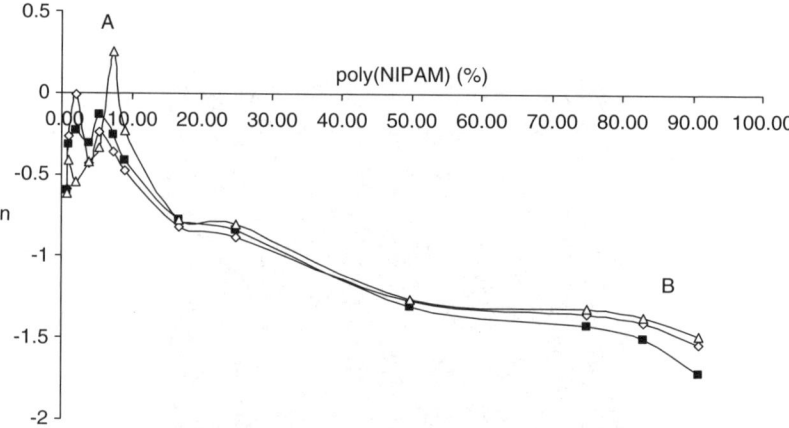

Figure 8.4   Heteroflocculation profile for poly(NIPAM) microgel (271 nm) and PS
latex (800 nm): (◇) 25 °C, (■) 50 °C and (△) redispersibility at 25 °C.

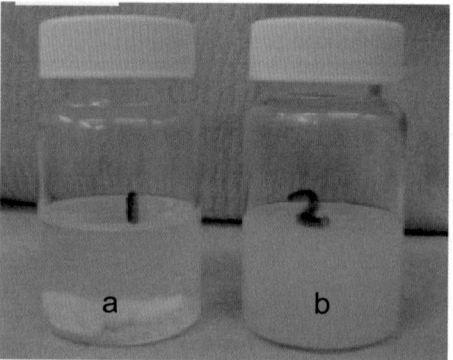

Figure 8.5   Visual observations made of samples at points A and B on the heteroflocculation profile (Figure 8.4): (a) flocculated dispersion at point A, and
(b) stable dispersion at point B.

interaction between the latex and the microgel results in the occurrence of a bridging process, where the smaller microgel particles form bridges between the larger PS latex particles (Figure 8.6). As the relative concentration of microgel particles to PS latex particles increases so does the potential for saturation adsorption. This results in the formation of stable heteroparticle complexes which in turn result in a stable dispersion ($n = -1.7$; Figure 8.4). Similar aggregation behaviour was observed by Hall *et al.*,[25] where the aggregation behaviour of oppositely charged microgel particles was investigated. It was found that aggregation between the particles occurs at the temperature below the VPTT of the microgels (20 °C) for mixtures where the cationic microgel is not in large excess and stable dispersions ($n = -2.44$) occur when the cationic microgel is in excess. There was no noticeable change in the stability of the dispersions at elevated temperatures (greater than the VPTT of poly(NIPAM)). The SEM analysis showed a good agreement with the aggregation profile determined by turbidimetric measurements. Similar phenomena were observed for mixtures of poly(NIPAM/AMA) (271 nm) microgel and 460 nm PS latex (PS 5) (Figure 8.7).

8.3.3.2 Cationic Poly(NIPAM) Microgel (271 nm) and PS Latex (100 nm). Mixed particle dispersions with low relative microgel to latex concentrations ($\leq 20\%$) showed stabilisation ($n = -2$ to -3; Figure 8.8) at room temperature (25 °C). SEM images confirm that this is a result of the formation of stable heteroparticles (Figure 8.9a). The latex and microgel particles look similar in size in the SEM images. This is perhaps not surprising as the microgel particles are in their collapsed state under vacuum. Saunders and Vincent noted similar TEM/light scattering size discrepancies. Heteroflocculation took place when the dispersion was heated above the VPTT of the microgel (50 °C). This is attributable to an increase in the surface charge density of the microgel particles in the collapsed state. This results in an electrostatic attraction between the heterofloccs through oppositely charged particles. The heteroflocculated sample did

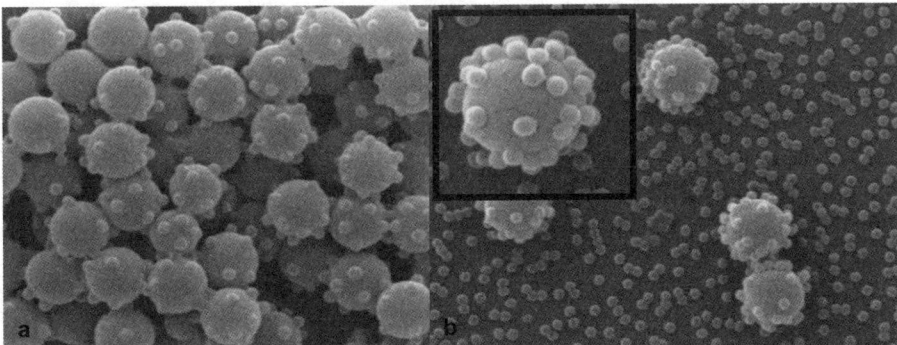

Figure 8.6   SEM images obtained for samples at points A and B on the heteroflocculation profile (Figure 8.4) for poly(NIPAM) microgel (271 nm) and PS latex (800 nm): (a) 7.5% poly(NIPAM), 92.5% PS latex; (b) 91% poly(NIPAM), 9% PS latex.

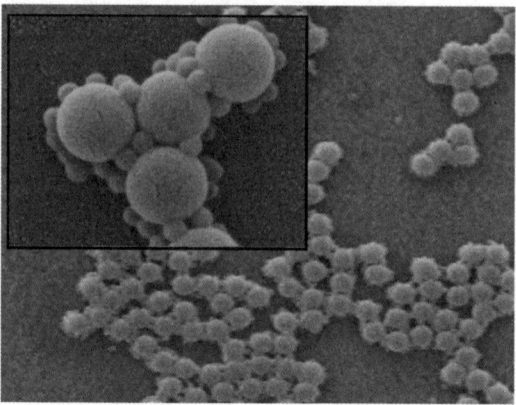

Figure 8.7   SEM images of 460 nm PS and 271 nm poly(NIPAM) 25% + 75% PS 5.

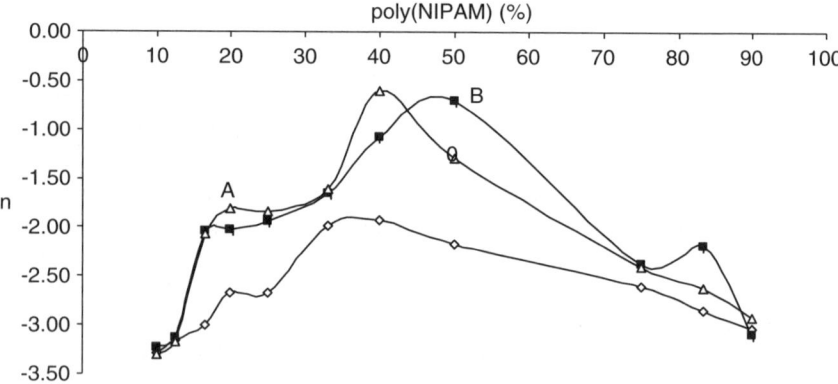

Figure 8.8   Heteroflocculation profile for poly(NIPAM) microgel (271 nm) and PS latex (100 nm): ($\Diamond$) 25 °C, ($\blacksquare$) 50 °C and ($\triangle$) redispersibility at 25 °C.

redisperse physically on shaking when the temperature was lowered below the VPTT. The dispersibility value does not indicate that the system was fully dispersed. This may be due to the hydrophobic entanglement of the polymer chains of the microgel with the surface of the PS latex. A visual observation of the mixtures indicated clearly under which sets of conditions the mixtures were aggregated or dispersed. An irreversible aggregation is observed at an equal concentration of both latex and microgel. The SEM image (Figure 8.9b) shows that the aggregated particles formed a network structure. This is a type of strong aggregation behaviour which potentially could be exploited in the field of enhanced oil recovery.[24] Islam *et al.* have reported similar behaviour in some cases; they observed heteroaggregation of oppositely charged poly(NIPAM) microgels

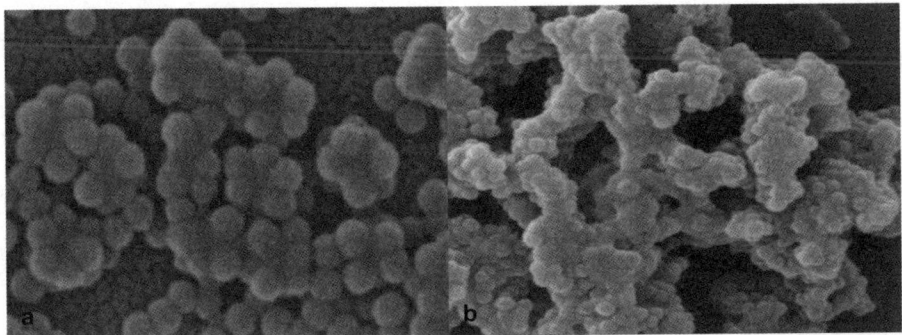

Figure 8.9 SEM images obtained for samples at points A and B on the hetero-flocculation profile (Figure 8.8) for poly(NIPAM) microgel (271 nm) and PS latex (100 nm): (a) 80% poly(NIPAM), 20% PS latex; (b) 50% poly(NIPAM), 50% PS latex.

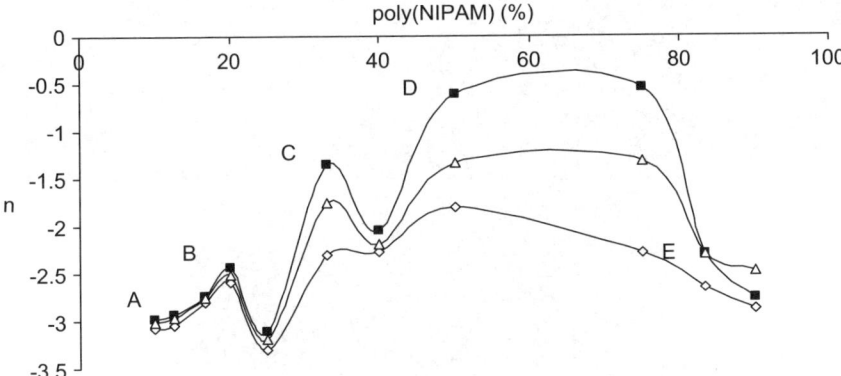

Figure 8.10 Heteroflocculation profile for poly(NIPAM) microgel (559 nm) and PS latex (100 nm): (◇) 25 °C, (■) 50 °C and (△) redispersibility at 25 °C.

and PS latex particles takes place when the mixed dispersion is heated above the VPTT of the microgel and the concentration of both microgel and latex in the dispersion is approximately equal.[15]

8.3.3.3 Anionic Poly(NIPAM) Microgel (559 nm) and PS Latex (100 nm). The heteroflocculation profile for this system is shown in Figure 8.10. Mixed particle dispersions with low relative microgel to latex concentrations (≤33%) showed stabilisation through the formation of stable heteroparticles, determined by SEM (Figure 8.11a–c). The heteroparticle size was found to vary as a function of particle concentration. In this case the stability of the system may be due to the net charge repulsion between the

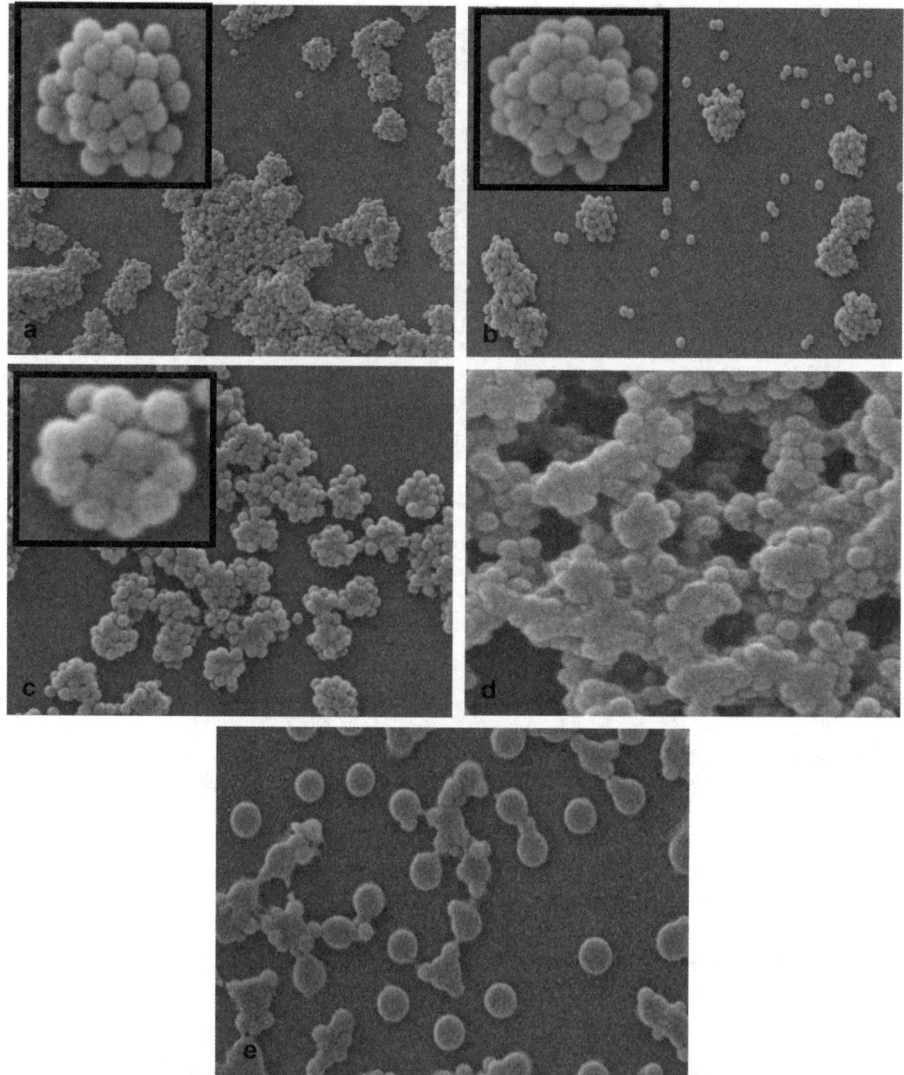

Figure 8.11   SEM images obtained for samples at points A, B, C, D and E on the
heteroflocculation profile (Figure 8.10) for poly(NIPAM) microgel
(559 nm) and PS latex (100 nm): (a) 12.5% poly(NIPAM), 87.5% PS
latex; (b) 20% poly(NIPAM), 80% PS latex; (c) 33% poly(NIPAM),
67% PS latex; (d) 50% poly(NIPAM), 50% PS latex; (e) 90% poly
(NIPAM), 10% PS latex.

macro-complexes, where an inner core of microgel is surrounded by negatively
charged PS particles.

The mixture of 33% poly(NIPAM) and 67% PS latex undergoes a
temperature-dependent reversible flocculation (Figure 8.12). At 25 °C a stable

Figure 8.12   Mixture containing 33% poly(NIPAM) microgel (559 nm) and 67% PS latex (100 nm) at: (a) 25 °C, stable; (b) 50 °C, flocculated; (c) redispersed at 25 °C, stable.

dispersion ($n = -2.3$) is observed. As the temperature is increased above the VPTT of the microgel, flocculation occurs ($n = -1.3$). This can be attributed to an increase in the electrostatic attraction between the microgel and latex particles. The flocculated samples redisperse (visibly) at 25 °C, *i.e.* below the VPTT of poly(NIPAM). However, the heteroflocculation profile determined by turbidimetric analysis of the mixtures indicates that the system is not fully dispersed at this point ($n = -1.8$). This could be attributed to the hydrophobic entanglement of the polymer chains of the microgel with the surface of the PS latex.

An aggregation ($n \approx -1.0$) is observed for a mixture of equal concentrations of both PS latex and poly(NIPAM) microgels at all temperatures. The SEM image (Figure 8.11d) shows that the aggregated particles have now formed a network structure. At relatively high concentration of microgel to latex ($\sim 90\%$ microgel), a stable dispersion was observed at both 25 and 50 °C ($n = -2.8$; Figure 8.11e).

8.4   Concluding Remarks

The heteroaggregation behaviour of cationic poly(NIPAM) microgels and anionic PS latex particles has been investigated with respect to particle size, surface charge, deformability of the particle, temperature and concentration. The physical state of aggregation was observed visually. The state of aggregation was studied by turbidimetric measurements together with SEM analysis.

The largest PS latex particles required the lowest microgel concentration to cause an aggregation. This may be attributable to an increase in the van der Waals attractive force between the particles with increasing latex particle size. It was found that the larger is the latex particle the less is the influence of temperature on the stability of the mixed particulate system.

The ionic interaction between oppositely charged deformable and non-deformable particles can be manipulated as a "switch on/switch off" heteroflocculation mechanism by changing the particle size, concentration and temperature of the constituent particles. The observed state of aggregation or stability is the result of a complex balance between diffusion effects (Brownian motion), repulsive and attractive forces as well as van der Waals forces—all of which are influenced by the environment of the particle, the particle size, the relative concentrations and surface charge density under a particular set of experimental conditions.

References

1. Th.F. Tadros and B. Vincent, in *Emulsions*, ed. P. Becher, Marcel Dekker, New York, 1982, p. 129.
2. B. Vincent, *Chem. Ind.*, 1980, 218.
3. B. Vincent, *Adv. Colloid Interface Sci.*, 1974, **4**, 193.
4. B. Vincent, in *Organic Coatings: Science and Technology*, ed. G.D. Parfitt and A. Patsis, Marcel Dekker, New York, 1983, p. 169.
5. D.W.J. Osmond, B. Vincent and F.A. Waite, *Colloid Polym. Sci.*, 1975, **253**, 676.
6. A. Fernández-Barbero and B. Vincent, *Phys. Rev. E*, 2000, **63**, 1.
7. B. Vincent, P.F. Luckham and F.A. Waite, *J. Colloid Interface Sci.*, 1980, **73**, 508.
8. G.J. Fleer, J.H.M.H. Scheutjens and B. Vincent, *Am. Chem. Soc. Symp. Ser.*, 1984, **240**, 281.
9. M.J. Snowden and B. Vincent, *J. Chem. Soc., Chem. Commun.*, 1992, 1103.
10. A. Routh and B. Vincent, *Langmuir*, 2002, **18**, 5366.
11. Th.F. Tadros, B. Vincent and C.A. Young, *Faraday Discuss. Chem. Soc.*, 1978, **74**, 337.
12. B. Vincent, C.A. Young and Th.F. Tadros, *J. Chem. Soc., Faraday Trans.*, 1980, **76**, 665.
13. M.J. Snowden, N. Marston and B. Vincent, *Colloids Polym.*, 1994, **272**, 1273.
14. B.R. Saunders and B. Vincent, *Adv. Colloid Interface Sci.*, 1999, **80**, 1.
15. A.M. Islam, B.Z. Chowdhry and M.J. Snowden, *Adv. Colloid Interface Sci.*, 1995, **62**, 109.
16. D.J. Shaw, *Colloid and Surface Chemistry*, Butterworth-Heinemann, 4th edn.
17. S.V. Vinogradov, *Curr. Pharm. Des.*, 2006, **12**, 4703.
18. V.C. Lopez, J. Hadgraft and M.J. Snowden, *Int. J. Pharm.*, 2005, **292**, 137.
19. S.V. Vinogradov, T.K. Bronich and A.V. Kabanov, *Adv. Drug Deliv. Rev.*, 2002, **54**, 135.
20. J.R. Retama, E.L. Cabarcos, D. Mecerreyes and B. Lopez-Ruiz, *Biosens. Bioelectron.*, 2004, **20**, 1111.
21. J.R. Retama, E.L. Cabarcos and B. Ruiz, *Talanta*, 2005, **68**, 99.
22. J. Kim, S. Nayak and L.A. Lyon, *J. Am. Chem. Soc.*, 2005, **127**, 9588.
23. P. Luckham, B. Vincent and Th.F. Tadros, *Colloids Surf.*, 1983, **6**, 101.
24. M.J. Snowden, B. Vincent and J.C. Morgan, *UK Pat.*, GB 226 2117A, 1993.
25. R.J. Hall, V.T. Pinkrah, B.Z. Chowdhry and M.J. Snowden, *Colloids Surf. A*, 2004, **233**, 25.

26. R.H. Pelton and P. Chibante, *Colloids Surf.*, 1986, **20**, 247.
27. J. Murray and M.J. Snowden, *Adv. Colloid Interface Sci.*, 1995, **54**, 91.
28. R. Pelton, *Adv. Colloid Interface Sci.*, 2000, **85**, 1.
29. M.J. Snowden, K. Kendal and R. Greenwood, *J. Eur. Ceram. Soc.*, 2000, **20**, 1707.
30. G.E. Morris, B. Vincent and M.J. Snowden, *J. Colloid Interface Sci.*, 1997, **190**, 198.
31. R. Hidalgo-Alvarez, A. Martin, A. Fernandez, D. Bastos, F. Martinez and F.J. de las Nieves, *Adv. Colloid Interface Sci.*, 1996, **1**, 67.
32. M. Rasa, A.P. Philipse and J.D. Meeldijk, *J. Colloid Interface Sci.*, 2004, **278**, 115.
33. S. Harley, D.W. Thompson and B. Vincent, *Colloids Surf.*, 1992, **62**, 163.
34. K. Makino, H. Kado and H. Ohshima, *Colloids Surf. B*, 2001, **20**, 347.
35. A. Fernandez-Nieves, A. Fernandez-Barbero and F.J. de las Nieves, *Langmuir*, 2001, **17**, 1841.
36. A.M. Islam, B.Z. Chowdhry and M.J. Snowden, *J. Phys. Chem.*, 1995, **99**, 39.
37. A.M. Puertas, A. Fernadez-Barbero and F.J. de las Nieves, *J. Chem. Phys.*, 2001, **115**, 5662.

Chapter 9

Surface Modification, Encapsulation and Coating: A Career Built on Graft

David Fairhurst

COLLOID CONSULTANTS LTD, CONGERS, NY 10920-1834, USA

Abstract

Colloid science can be used to provide significant benefits in the design and manufacture of personal care, cosmetic, pharmaceutical and medical products and processes. It has been called the midwife of invention; somewhere in the function, formulation and processing necessary to achieve new and improved products the influence of colloid science, though sometimes hidden from the untrained eye, will be felt. Brian Vincent has contributed much to the fundamentals of this science, in particular pushing forward the theoretical boundaries. Paraphrasing Nobel laureate Richard Smalley, if colloid science research is the "garden of the physical sciences", then Brian also has been a creative master gardener harvesting solutions to many a thorny problem (no pun intended). And, like a good artisan, he has passed on his skills to those who have been fortunate enough to have been called his students. That he is an excellent teacher is evident from the many eminent scientists worldwide continuing his legacy.

Trained in this discipline, I have continued to work for nearly forty years, not as a scientist or as an academic but as an instrument designer, cosmetics formulator and salesperson. Unlike many in academe and industry I have not stayed "in one place" but have moved as new opportunities and new challenges arose. In so doing I became a "jack-of-all-trades" but have had the good fortune of never having been bored. My contribution to this book celebrating the career of Brian Vincent is thus as an eminent businessman who has dabbled in colloid science.

New Frontiers in Colloid Science: A Celebration of the Career of Brian Vincent
Edited by Simon Biggs, Terence Cosgrove and Peter Dowding
© The Royal Society of Chemistry 2008

My contact with Brian dates back to 1968, and so this chapter will first chart the early associations with him and then focus on a few examples of how I have used some practical facets of colloid science such as surface modification, encapsulation and coating to build a successful and, financially, extremely rewarding business. I will end with a couple of examples of how the application of colloid science might help in the fight against HIV/AIDS. Throughout Brian's career I have been privileged to have him as a friend and to be able to call on his vast experience in the field of colloid and interface science. Discussions with him were always lively and instructive; the help and advice he has given me over the years has been invaluable. He is *the* epitome of the truly eminent scientist.

9.1   The Early Years: Colloid Science, Zeta Potential and Polymers

My first contact with Brian was in 1968, when as a lowly graduate student at the Liverpool College of Technology (now Liverpool John Moores University) I had been given the task of helping to organize a symposium on electrophoresis and electrokinetic phenomena. My PhD thesis (1968) would be titled "Electrokinetic phenomena at the silver halide–solution interface". Pertinent to photographic processes, the work was supported by a grant from Ilford Ltd. The studies required measurement of electrophoretic mobility (zeta potential) of silver halide sols and so I designed a new microelectrophoresis instrument, the prototype of which (Figure 9.1) was commercialized and sold as the Rank Bros. MK II microelectrophoresis apparatus. It is clear that the seeds of my transition to the "dark side" were being sowed even then.

The electrophoretic technique had been a major tool for studying the electrical double layer around colloidal particles. At that time, it was generally assumed that the primary sources of error arose from focusing (*ca.* 3%) and Brownian motion (*ca.* 3%). I presented a paper showing how serious both electrode polarization and Joule heating effects could be in obtaining precise electrophoretic mobility measurements. These effects could be minimized by the use of a thin-walled microelectrophoresis cell and electrodes of Pt-black or combination Pd/Mo electrodes. Years later I would put this to good effect in the design of cells for two other commercial zeta potential instruments—the PenKem Model 500/501 and the PenKem System3000—and finally in the design of a cell to eliminate the problem of electroosmosis (Brookhaven ZetaPlus).

My supervisor was the late Prof. Alec Smith, who was a friend of Prof. Ron Ottewill. In part, Ron had been invited in order to meet me as he was scheduled to be my thesis examiner. At the symposium I mentioned to Ron about some results I had on adsorption of alcohols onto silver iodide. It turned out that part of Brian's thesis work was concerned about adsorption of alcohols onto polystyrene and Brian was kind enough to send me some of his data. We also had a chance to "compare and contrast" the personalities of our respective

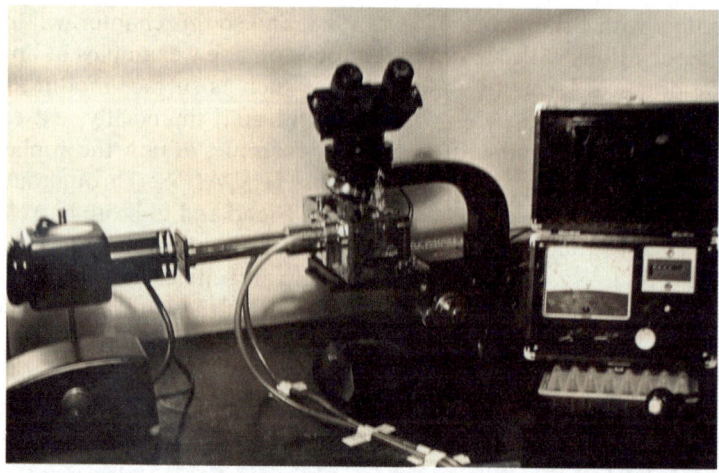

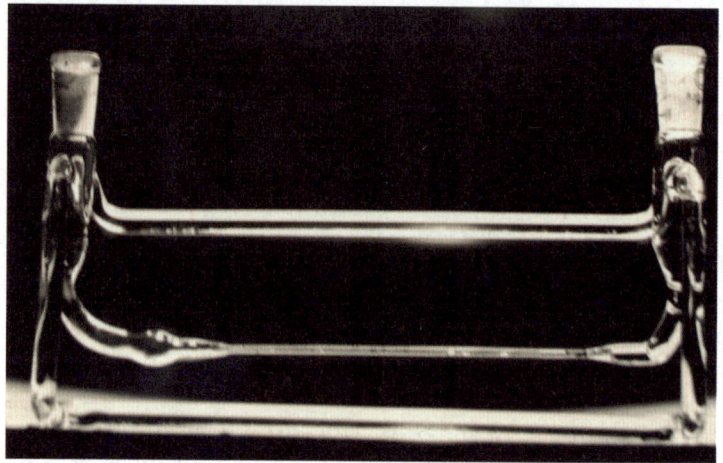

Figure 9.1 Microelectrophoresis apparatus (top) and thin-walled Van Gils type cell (bottom).

supervisors! That contact cemented a friendship and was the first of many in-formal discussions and a couple of collaborative projects that have occurred over the last forty years. So, to say that he has had an influence on my career is an understatement.

After obtaining my PhD, I was awarded a one-year Unilever Research Fellowship, at Unilever Research Laboratories, Port Sunlight, UK (with Dr Colin Smart), to study the dispersion properties of builders in mixed water/solvent systems. This was followed by two years as a visiting research associate with Professors A. C. Zettlemoyer and F. J. Micale at Lehigh University, USA, investigating dewatering phenomena in activated sludges for the US Environmental Protection Agency. In addition, I worked on ink formulation at the

National Printing Ink Research Institute (part of the Center for Surface and Coatings Research, Sinclair Laboratory)

Returning to Liverpool Polytechnic as a lecturer in physical chemistry, I started to get interested in surface modification; adsorption of the low molecular weight surfactant octylmethylsulfoxide onto both silver iodide and graphitized carbon aqueous sols has both a stabilizing and sensitizing action. Both effects could be sufficiently explained by electrostatic effects and dispersion attraction with only a modest thickness (*ca.* 1 nm) of adsorbed layer, *i.e.* without recourse to the postulate of extensive solvation layers.[1] In 1973, I organized a symposium on light scattering methods for particle counting and the study of flocculation kinetics and I invited Brian (by then a newly minted lecturer at Bristol University) who gave a paper on "Equilibrium aspects of reversible flocculation".

It was apparent then that he was destined to be an outstanding lecturer and teacher. Brian's understanding of light scattering also helped me to master some of the complexities of that technique and would stand me in good stead when later (at Brookhaven Instruments Corp.) I was responsible for the sale of particle size analyzers based on laser light scattering and in the development of a zeta potential instrument based on electrophoretic light scattering.

However, it became clear to me that I was not going to make it as an academic and so, in 1974, I joined the Civil Service! Specifically, I joined the Protection Division of the (MOD) Chemical Defence Establishment at Porton Down. At CDE, I was responsible for Special Projects R&D, with emphasis on protective systems against chemical warfare agents, for both the UK and NATO armed forces. While there, I collaborated with Dr Mike Wilkinson on the production and characterization of (emulsifier-free) polymer colloids.[2,3] At the time, there was a lot of academic interest in polystyrene (PS) latices not just for use as "model" colloids but also as reference materials to validate particle size and zeta potential instruments and for use as indicator particles in the diagnosis of pathological conditions. The original preparative "emulsifier-free" method was described by Matsumoto and Ochi[4] and we used variations in this method to produce a wide range of different polymer colloid systems from polystyrene to poly(butyl methacrylate).

My main interest, however, was to improve performance and production of the nuclear, biological and chemical (NBC) protective suit used by the armed forces. This comprised an inner layer of a non-woven fabric to which had been bonded coconutshell charcoal as a high-efficiency adsorbent (typically *ca.* $1000 \, m^2 \, g^{-1}$ total adsorptive capacity) for chemical warfare (CW) agents. Coconut shell charcoal is unique in that the pore size distribution is fairly uniform (Figure 9.2).

The charcoal was bonded to a non-woven cloth using a commercial industrial polymer (acrylate-type) latex. An aqueous co-slurry of charcoal and latex was simply sprayed onto the cloth surface and then cured producing a flexible adhesive bond between the charcoal and the cloth. Two problems were manifest in manufacturing. The first was that because of the broad particle size distribution of the latex, penetration into the porous structure of the charcoal was unavoidable. This led to a dramatic reduction in the dynamic adsorption

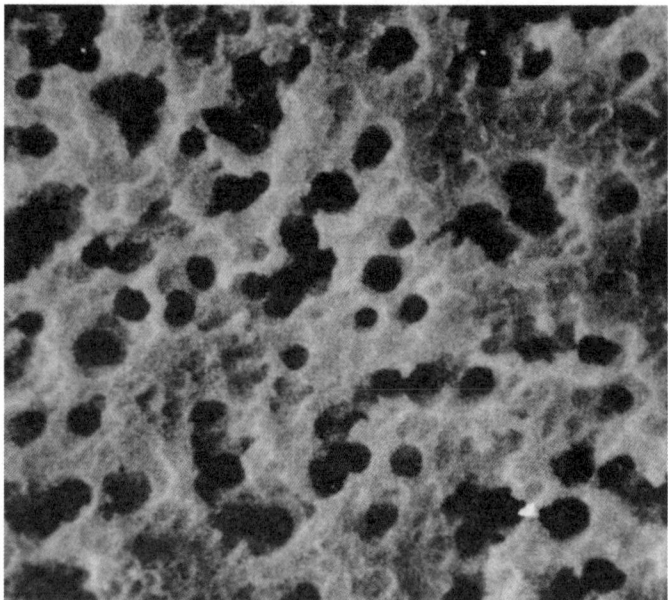

Figure 9.2   Electron micrograph of coconut shell charcoal.

rate of CW agents (although the equilibrium adsorption was only marginally affected) rendering the material essentially useless as a protective barrier. Thus, I reasoned that using monodisperse latices of a mean size larger than the largest pore size in the charcoal should eliminate the penetration problem. The difficulty was making them big enough. We needed to make them bigger than 1 μm. Sadly, I did not make the intuitive leap like Ugelstad! The second problem was that the co-slurry tended to heteroaggregate particularly at the spray nozzle. Manipulation of the surface charge of polymer latex through different initiators and/or subsequent adsorption of charge modifying agents reduced this to manageable proportions.

In the mid-1970s, Brian had become interested in the preparation of dispersions of *neutral* particles carrying a well-defined polymer layer in order to test the theories of steric interactions.[5,6] It was known that such systems could be prepared by non-aqueous, free radical dispersion polymerization techniques[7] and stabilized by AB block, ABA block or comb graft copolymers. However, a disadvantage of all polymer systems was the potential for swelling by the solvent. It was then in late 1977 during one of our numerous "chats", when I was picking Brian's brains about manipulation of surface charge of polymer colloids, Brian suggested that stable, non-aqueous dispersions of spherical, monodisperse, non-swelling silica particles might be prepared using anionic polymerization techniques to terminally graft polymer chains to the particle surface. This struck a chord because I had just been asked to look at ways in which a self-decontaminating paint topcoat might be formulated to provide protection for armoured vehicles. CW agents were known to permeate/diffuse through polymer film such as that used in (oil-based) paint.

Also, it was known that MgO/Mg(OH)$_2$ could hydrolyze CW agents. Thus, if such particles could be incorporated into a paint formulation they might provide some level of protection. Accordingly, I arranged for a CASE award to study the feasibility of preparing such dispersions; the graduate student was Keith Bridger. The idea was first to just use silica particles as a test substrate; they could also be incorporated into a paint film to provide mechanical integrity similar to the use of syloids in varnishes. The silica particles were prepared using the Stöber method[8] and the grafting reaction was that proposed by Papirer and Nguyen.[9]

Keith spent six months at Porton Down building an updated version of the Bristol high-vacuum rig needed to carry out the anionic polymerization (Figures 9.3–9.5). The success of this work[10] led to these PS-*g*-SiO$_2$ dispersions being used by Brian's laboratory to test the steric interaction theories and the

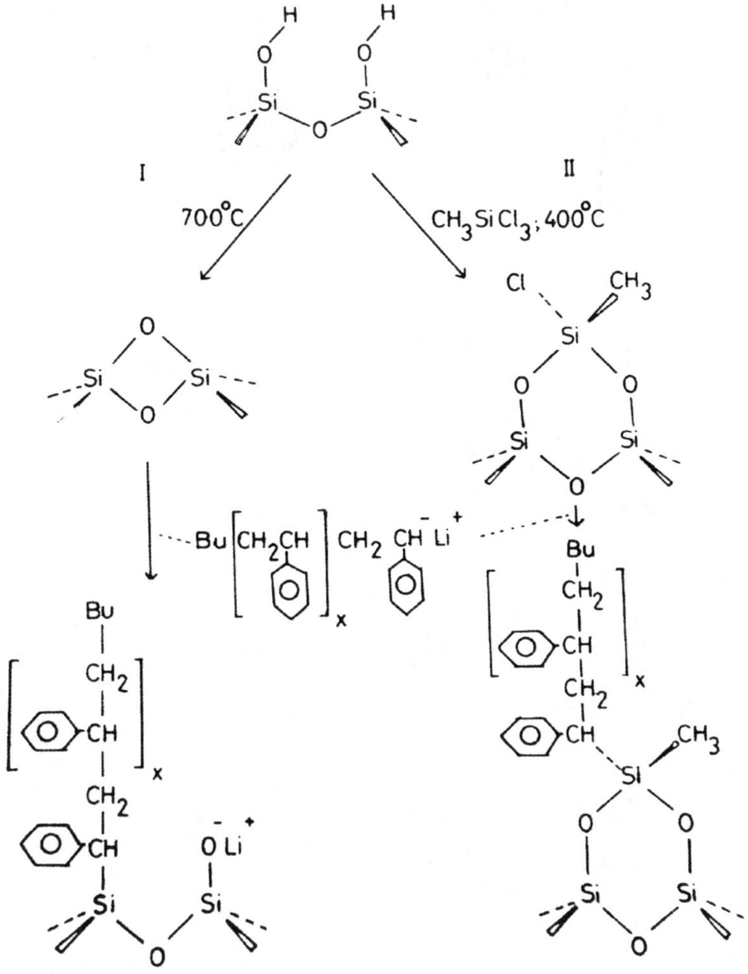

Figure 9.3 Schematic of (anionic) grafting routes for polystyrene onto silica particles.

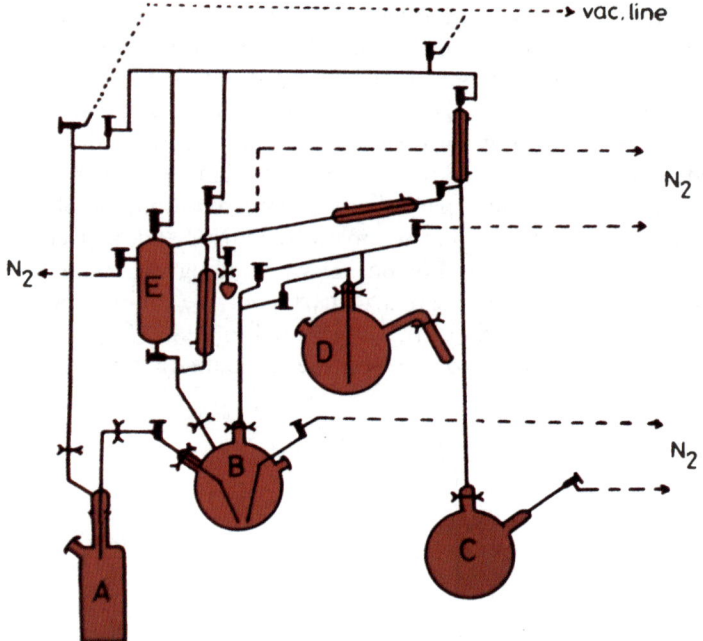

Figure 9.4   Anionic polymerization rig. (Note that vessel D is only used for grafting polydimethylsiloxane.)

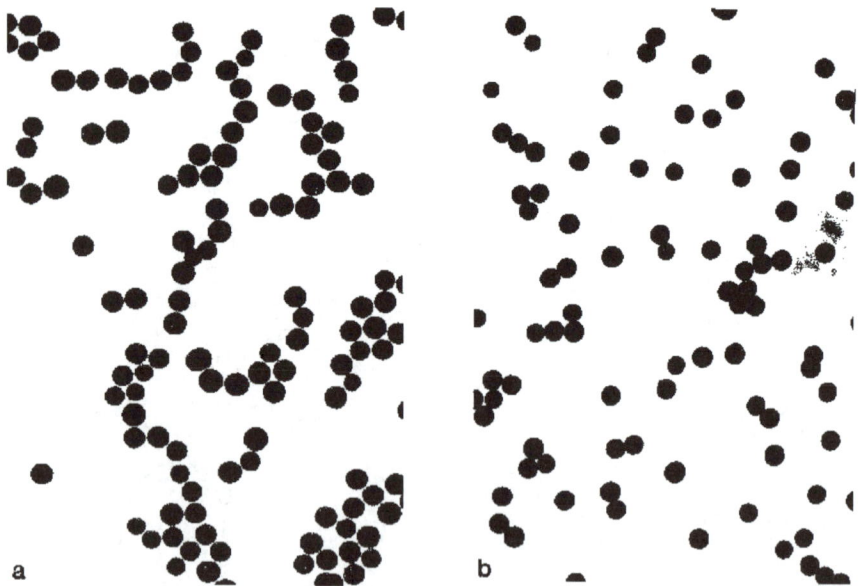

Figure 9.5   Electron micrographs of silica particles (a) before and (b) after anionic grafting of polystyrene. Note there is no apparent change in morphology and the particles are present as singlets. Average particle size is 350 nm.

process being extended to grafting polydimethylsiloxane and polyethylene to silica and other inorganic particles.

At the time, anionic polymerization was cutting-edge stuff but it has now been superseded by controlled radical polymerization[11] because of that technique's numerous advantages: controlled radical polymerization is solvent-free, a one-step room temperature synthesis and is much less expensive.

To celebrate the conclusion of the CASE award, I joined Brian, Keith and other graduate students on a (Bristol University) bus trip to the Netherlands; the goal was Wageningen and the 1979 Colloid Stability Symposium. It was during this trip that the "Incident at Jülich" occurred where Brian managed to get me arrested and put in prison!

Sadly, I was not permitted to carry the paint project forward as, unfortunately, changes were in the air. A Defence Review, ordered by the then British prime minister Margaret Thatcher resulted, just after Keith Bridger left, in a reorganization at Porton Down; it was, in reality, the beginning of the dismantling of the scientific arm of the Civil Service. One immediate consequence was that the Protection Division was closed down and I, in typical army fashion, was reassigned to the Detection and Analysis Division! So, it was time for a career change.

9.2   The Dark Side: Black Boxes, Sunscreens and Hope-in-a-Bottle

Since the overwhelming majority of manufactured products involve, either in the final state or at some stage of their production, suspensions of particulate materials, emulsion droplets or air bubbles, the need for instrumentation to characterize fundamental parameters such as particle size and zeta potential is obvious. But, back then, virtually all such instrumentation was "home made". That which was commercially available tended to be manual and time-consuming. I, amongst others, recognized that there was a need for standardized instruments that provided fast, automated measurements. With the advent of the PC the opportunity arose. I succumbed to the almighty dollar and, in 1980, moved to the USA and spent the best part of the next ten years as an odious salesperson and purveyor of those black boxes first with PenKem Inc. and then with Brookhaven Instruments Corp.

During this "interregnum" Brian, once again, provided me with an opportunity but this time it would be a commercial venture. By the very late 1980s nearly every R&D laboratory had purchased a zeta potential instrument based on electrophoretic light scattering (ELS); the technique was fast, reliable and precise.[12] However, ELS was severely limited in its ability to measure precisely the very low electrophoretic mobilities found, for example, with dispersions in non-polar media where the dielectric constant is low; such measurements are, experimentally, more difficult than the corresponding measurement in aqueous media.[13] The stabilization and behaviour of such suspensions are of considerable interest: systems of this type are widely used in paint and coatings, in lubrication technology, in pharmaceutical, agricultural and cosmetic formulations, in reprographic applications and in the development of high-performance ceramics. In order to measure sterically stabilized suspensions in non-aqueous media, Brian's graduate student, John Miller, had modified (both the hardware and software) a standard zeta potential instrument.[14] John used what is called phase modulation; a suggestion originally made by Klaus Schätzel.[15] In so doing he demonstrated, using a copolymer of dimethylaminoethyl methacrylate/methyl methacrylate-g-SiO$_2$ in 1,4-dioxan, that this technique was capable of determining electrophoretic mobilities down to $10^{-12}\,\mathrm{m^2\,V^{-1}\,s^{-1}}$ (Table 9.1)—this would correspond to only one or two charge sites/groups per particle!

On one his many visits to the USA, Brian and I met and, upon hearing about John's work, I immediately arranged for a collaborative project with Brian to develop a commercial version of John's instrument. John, who had decamped to a postdoctoral position with Harry Ploehn (then at Texas A&M University), was hired as a "consultant". A number of very significant improvements were made by Fraser McNeill-Watson, including the use of digital signal processing, a reference beam configuration and a very high frequency field. Brian's laboratory was the beta-test site for what became the Brookhaven ZetaPALS.[16]

Non-aqueous systems are not the only ones that pose a challenge. Aqueous systems can be just as difficult. For example, severe difficulties arise in obtaining reliable data in environmental applications such as brine and sea water, as measurements above 0.1 M cannot be performed reliably with conventional ELS. Using phase analysis light scattering (PALS), however, it is possible to obtain data in highly saline solutions (Table 9.2) and even saturated salt (*ca.* 4 M).

Table 9.1 Electrophoretic mobility $(10^{-11}\,\mathrm{m^2\,s^{-1}\,V^{-1}})$ of poly(dimethyl-aminoethyl methacrylate)-*co*-poly(methyl methacrylate)-*g*-SiO$_2$ in 1,4-dioxan using the PALS method. Note the exceptional low value for the mobility and good repeatability and that the PALS method only gives an average value.

Four independent measurements: 5.21, 5.54, 4.48, 4.41
The average mobility of 4.89 ($\pm$0.6) is three orders of magnitude smaller than typical
   aqueous values

Table 9.2 Electrophoretic mobility of BCR66 quartz as a function of KCl (aq) concentration. Note the mobility in 4 M KCl is still negative.

KCl (aq) concentration (M)	*Mobility $(10^{-8}\,m^2\,V^{-1}\,s^{-1})$*
0.0001	−4.30
0.001	−4.56
0.01	−3.5
0.1	−2.1
0.5	−1.75
1.0	−1.33
2.0	−0.73
4.0	−0.01

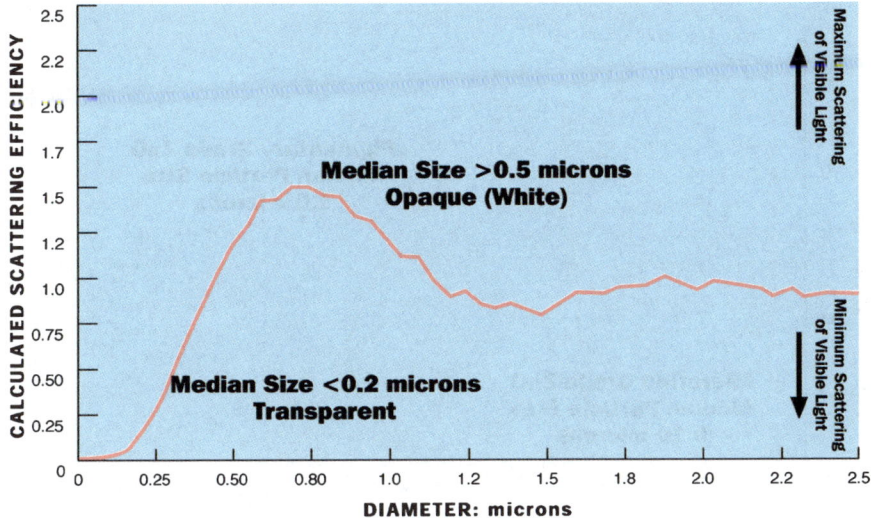

Figure 9.6 Scattering efficiency of ZnO to white light. Calculations according to Mie theory.

And then a serendipitous meeting with a paediatrician opened up an avenue to a new venture. Dr Mark Mitchnick had a practice in East Hampton, NY—a famous watering hole for the "glitterati" of New York, but also an extremely popular summer holiday retreat. He was used to treating babies and young children for the effects of overexposure to the sun. He knew that the zinc oxide (ZnO) cream used to treat baby nappy rash was a very effective sunblock. Unfortunately, none of the parents of his patients would allow this to be used because the product was extremely whitening when applied to the skin (it was also thick and greasy since it was an ointment base). The question he posed was whether it was possible to make a cream or lotion with a ZnO that was "transparent". I, in my colloid science mode, replied that it was a "piece of cake". The cream was white because (a) the ZnO concentration was so great (*ca.* 40%) and (b) the particle size was too large (*ca.* 800 nm).

Thus, all one had to do was to make the ZnO particles smaller than about 200 nm[17] (Figure 9.6) so that the human eye could not detect them and use a concentration less than about 10%. At that size and concentration the particles would not be detected by touch either so a formulated lotion or cream would be aesthetically pleasing. On the basis of this Mitchnick, his brother and a friend raised $1 million! I then sold my interest in Brookhaven Instruments and together we created a company called sunSmart Inc. Of course, actually making the ZnO was another matter but we were fortunate enough to joint-venture with Zinc Corporation of America who manufactured the material—we called it ZCote®, microfine zinc oxide—exclusively under license for us (Figure 9.7).

Advances in the understanding of photoageing and aggressive skin cancers had highlighted the need for sunscreen agents that not only prevent sunburn but

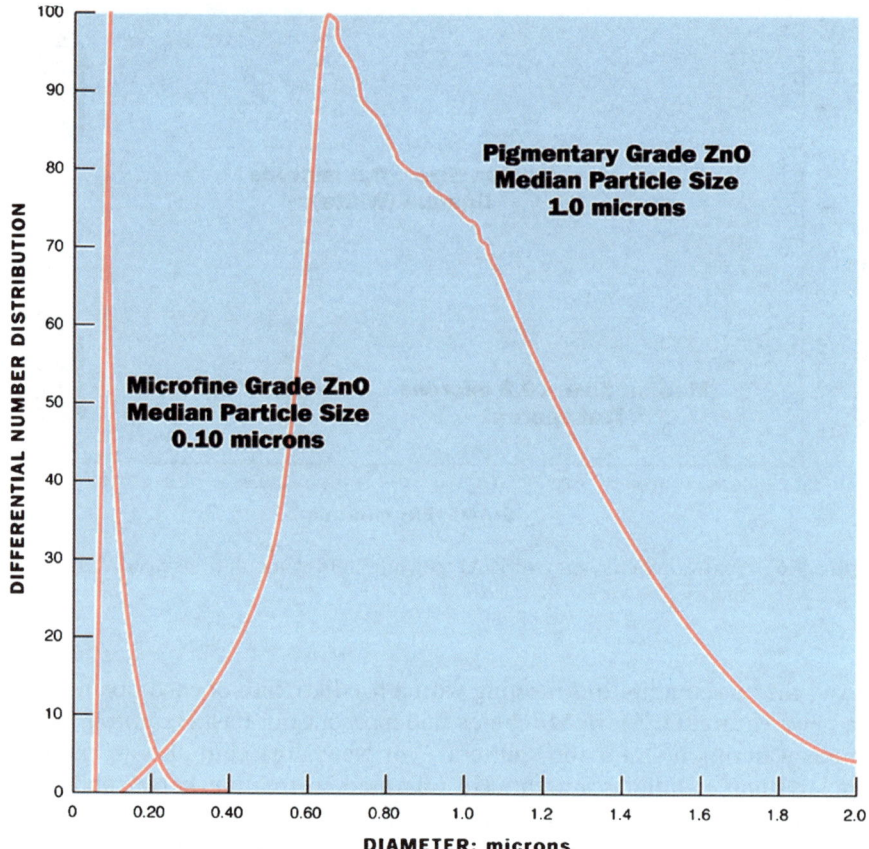

Figure 9.7 Particle size distribution of microfine zinc oxide compared with pigmentary grade material.

also block UVA irradiation (320–400 nm). UVA penetrates deeply into the dermis (Figure 9.8) and can promote immunosuppression[18] which is highly associated in humans with the development of skin cancers[19] and it can, through generation of singlet oxygen, produce metalloproteinases that can destroy the connective tissue of skin.[20] Alone among available sunscreen actives, ZnO covers the widest UV range (Figure 9.9); indeed, it also absorbs in the infrared.[21] In theory, it is possible to make a broad-spectrum (UVA plus UVB) sunscreen using only ZnO.

ZnO is an inorganic "particulate" or "physical" sunscreen active and it presents formulation difficulties unlike those of traditional soluble organic "chemical" sunscreen actives. Sunscreen formulations are essentially oil-in-water (O/W) or water-in-oil systems (W/O) in which the active is typically dispersed in the oil phase. Generally, O/W emulsions are used for moisturizer/daily wear products and W/O for beach use (because such formulations can be made water

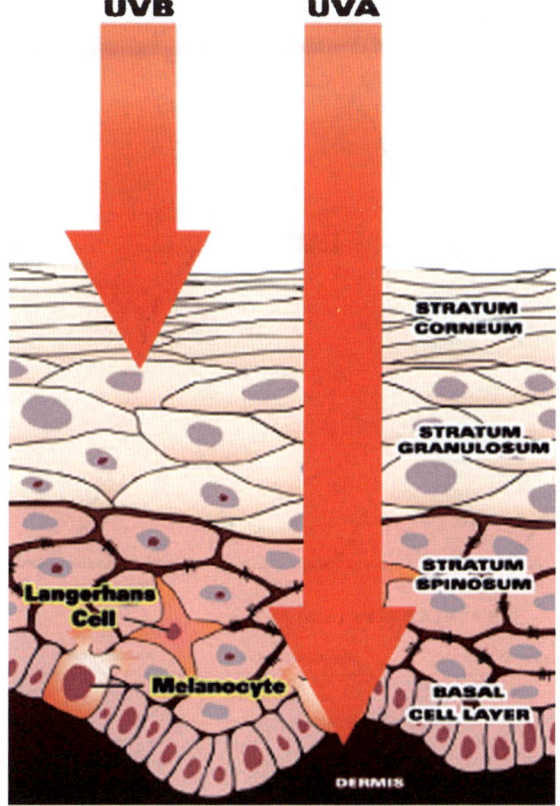

UVB and UVA both cause catastrophic and carcinogenic events in the skin **but** they do so in very different ways

Figure 9.8 UV radiation and the skin.

resistant). Unfortunately, ZnO is very hydrophilic and proper wetting and dispersion is needed to avoid agglomeration and also to keep it in the oil phase.[22,23] ZnO cannot be used in the water phase because below pH = 6 it is increasingly soluble and this can affect the emulsion stability. It has a small but finite solubility in neutral water and this manifests itself as an increase in pH owing to the formation of alkaline zinc complexes. This is turn means that the use of, for example, stearic acid and carbomers (two excipients widely used by cosmetic and pharmaceutical formulators) must be avoided because of unwanted reactivity with the complexes.

Harking back to my Porton Down days, the obvious answer was to coat the ZnO particles by grafting a reactive silane to the surface. Again, this was easier said than done for a number of reasons. First, in the USA, sunscreen actives are regulated as drugs by the US Food and Drug Administration (FDA). This means that they must meet the USP for purity. Second, the activity depends on the ZnO concentration; any coating would need to be less than about 2% so as not to adversely impact the UV absorption efficacy. Third, the complete

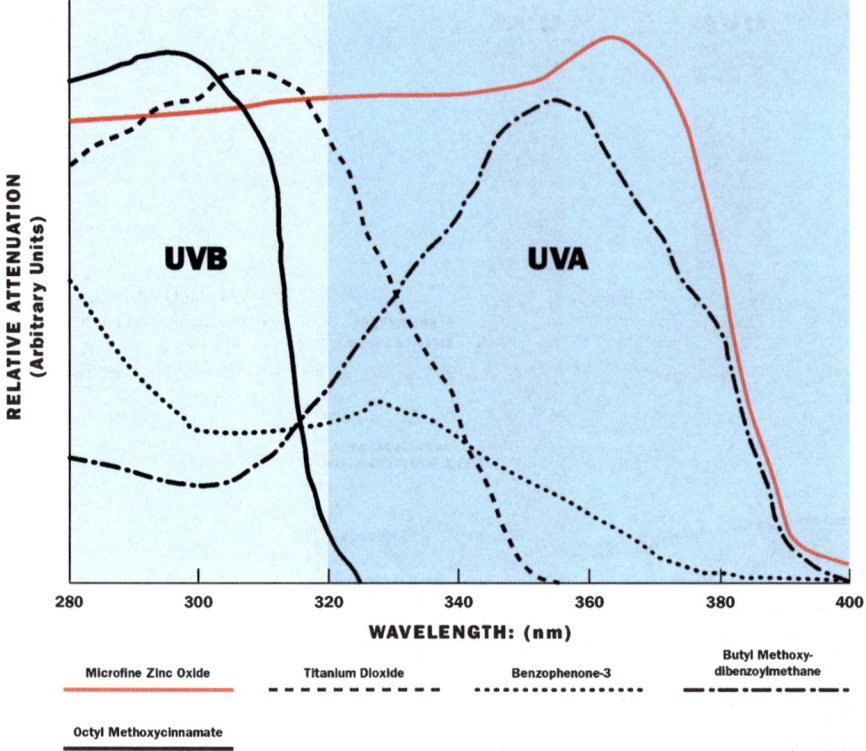

Figure 9.9   UV attenuation of various sunscreen actives.

commercial manufacturing process had to be very simple and also could not significantly add to the cost. Working with Dr Tony O'Lenick (Siltech LLC, Galen, GA, USA), I came up with a proprietary reactive silane that could be applied using a simple V-blender at room temperature. After reaction, this reactive silane created a coating similar to dimethicone in its chemical composition. At <2% the coating was robust and exceptionally hydrophobic. The result was a product called ZCoteHP1®, which proved to be very successful since it made the job of the cosmetic formulator much easier.

But the coating also provided one unexpected benefit. With the recognized importance of UVA radiation one open issue was the photostability and photoreactivity of actives that attenuate in the UVA region.[24–26] A key reason to use a sunscreen is to "sacrifice" the extrinsic molecule (the organic active) to spare the body's own chromophores, *i.e.* to prevent photodegradation of the living tissue below them; melanin is known to produce free radicals under irradiation.[27] The photostability of ZnO is thoroughly documented[21,28] and has not been seriously questioned. However, the FDA had become increasingly concerned about the photoreactivity of TiO_2 (another inorganic particulate sunscreen active) and ZnO got hit by the shrapnel. Now, photoreactivity can be evaluated using a method based on the photocatalytic oxidation of iso-propanol, a well-established and recognized procedure.[29,30]

Using this technique, I was able to demonstrate that the ZCote® was essentially unreactive and that ZCoteHP1® was inherently non-photoreactive under any circumstances (Figure 9.10). In addition, I showed that ZnO when used in combination with commonly used organic sunscreens slows their degradation on being subjected to solar simulator irradiation (Table 9.3). These results were presented to the FDA during the discussion and comment phase prior to publication of the FDA Final Monograph on Sunscreen Drug Products for Over-the-Counter Human Use.[31]

The result of all this was that by 1999 we were able to sell over 500 000 lb/annum of microfine ZnO for use in cosmetic and daily wear products and in 2000 we sold all the intellectual property and manufacturing rights to ZCote®/ZCoteHP1® to BASF, for a very tidy sum! And, in 2003, ZCote® was voted by Forbes magazine as one of the top ten nanotechnology products for that year.[32]

With the shareholders all handsomely paid off, what to do next? We had originally also set up a subsidiary company called Particle Sciences Inc. (PSI) to be the "research arm" of sunSmart (it provided a number of fiscal and tax

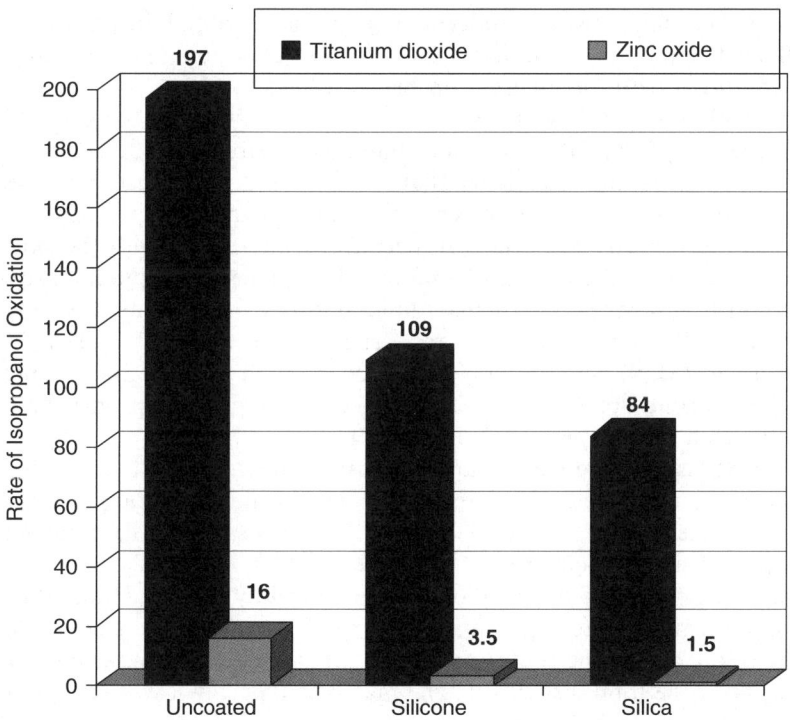

Figure 9.10   Relative photoreactivity of coated and uncoated microfine grades of zinc oxide and titanium dioxide.

Table 9.3 Photoprotective effect of ZnO to Octinoxate. Irradiation was *via* a solar simulator ($10\,J\,cm^{-2}$ is equivalent to a one-hour exposure in June at 40° north latitude).

Formulation	Octinoxate remaining after irradiation (%)		
	$10\,J\,cm^{-2}$	$20\,J\,cm^{-2}$	$30\,J\,cm^{-2}$
7.5% Octinoxate	85.1	80.7	78.1
7.5% Octinoxate + 10% ZnO	87.8	85.8	83.8

advantages) and this remained after the sell-off to BASF. Prior to the introduction of the "physical" sunscreen actives, the organic sunscreens had been the mainstay of photoprotection. Because they are applied topically to the skin in relatively high concentrations (up to 25% total) and they eventually systemically absorb through the stratum corneum,[33] contact sensitization can occur. Because they absorb radiation they also have the potential to cause photosensitization;[34] photosensitization reactions can be either photoallergic or phototoxic. There are also a number of very significant diseases—the photodermatoses—that render those afflicted very sensitive to sun exposure.[35] These include polymorphous light eruption, the porphyrias, chronic actinic dermatitis, lupus erythematosus, xeroderma pigmentosum and albinism. Individuals so afflicted require the continued use of sunscreens potentially complicating their condition with the inherent problems of sensitization.

Encapsulation has long provided an invaluable tool to the cosmetics or pharmaceutical formulator, by imparting great flexibility in the choice of delivery mechanisms and excipients that can be used. It occurred to me that encapsulating the organic sunscreen active in a benign matrix would effectively turn them into particulates, in turn decreasing interaction with the skin (and thus irritant/allergic sensitization potential) and hence it should provide a significant health benefit. Particulate lipid matrices, in the form of lipid pellets, were originally developed for oral drug administration[36] and solid-lipid nanospheres (SLN) were initially produced by high-pressure homogenization of melted or solid lipids.[37] At PSI, we devised a low-pressure proprietary melt–emulsify–chill (MEC) process[38] (Figure 9.11).

These SLNs are produced as aqueous suspensions. As such they are designed to stay in the water phase of an emulsion. The particle sizes range from 0.1 to 1 μm; in the example shown in Figure 9.12 the mean particle diameters are between 300 nm and 400 nm.

The particle size and distribution are controlled by the rate of initial mixing, degree of shear and rate of cooling. These particles are a homogeneous mixture of matrix and active; the active can be present at concentrations as high as 30% by weight of the total dispersion. The matrix is typically a wax (or mixture of waxes) but may be an organic polymer or a silicone. A list of the controllable particle parameters is given in Table 9.4. In principle, the MEC process allows systems to be custom-designed to fit specific applications.

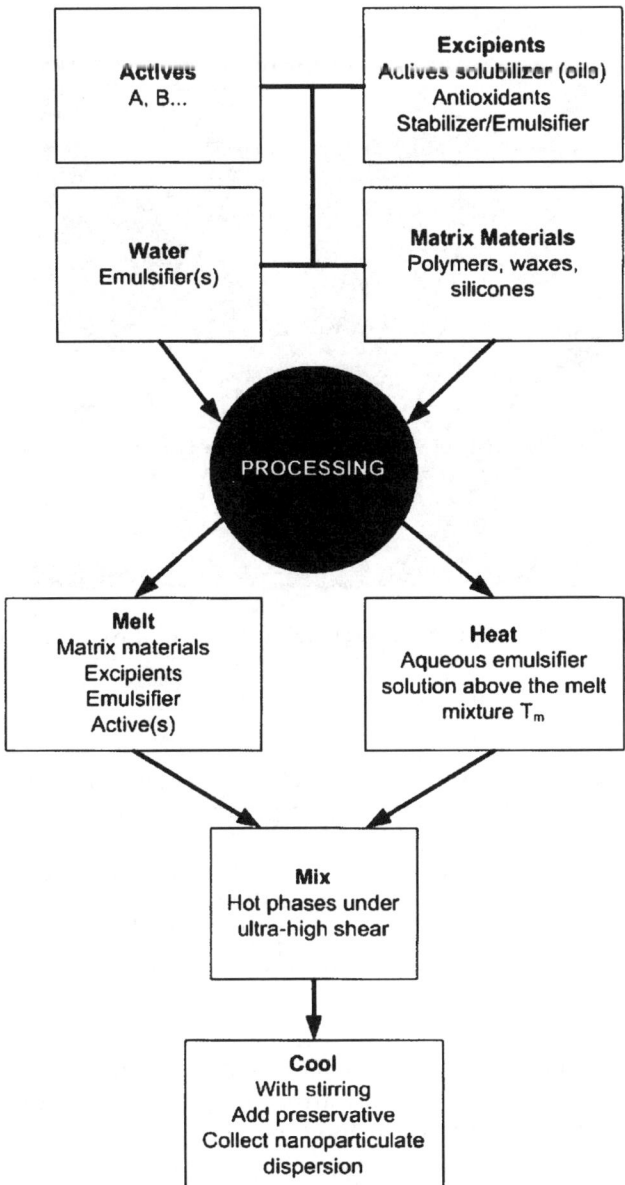

Figure 9.11   Schematic of the melt–emulsify–chill process.

I formulated three SLN suspensions containing organic sunscreens (Octi-noxate, Octinoxate plus Benzophenone-3 and Octinoxate plus Octocrylene) into a simple O/W moisturizing sun product and then measured the *in vitro* sun protection factor (SPF) values.[39] Three similar sun products, but containing non-encapsulated active, were used as controls. All of the formulations were

1 μm

Figure 9.12 Scanning electron micrograph of a typical SLN. Note that the suspensions are polydisperse.

Table 9.4 List of the controllable particle parameters in the MEC process.

Bulk physical characteristics	*Surface chemical properties*
Melting point, hardness, density, solubility	Surface charge (anionic, non-ionic, cationic)
Particle size and distribution	Hydrophobic/hydrophilic balance
Bioadhesion/tack	Polarity

sent for *in vivo* measurement on human subjects (AMA Laboratories, New City, NY, USA).[31] The results (summarized in Table 9.5) clearly demonstrate a substantial increase in SPF when using encapsulated sunscreen active compared with the control products containing regular sunscreen active at the same levels. Since sunscreen actives tend to be used at a high concentration (often >20% of the formulation) and they are usually the most expensive components, the cost/benefit ratio of using sunCaps® (as I termed them) was obvious.

For the encapsulated sunscreens, since the primary goal was to minimize bioavailability (penetration) of the active, a matrix was chosen that was chemically and physically compatible with the sunscreen active (*e.g.* Octinoxate); thermodynamically, there is then a low incentive for the active to diffuse out of the matrix. To test this hypothesis, we encapsulated anthralin using the

Table 9.5   Increase in sun protection factor obtained with encapsulation of organic sunscreens.

Sunscreen	In vitro control (%)	In vitro SLN (%)	Increase (%)	In vivo control (%)	In vivo SLN (%)	Increase (%)
5% Octinoxate	6.5	19.8	300	5.9	16.7	283
5% Octinoxate/ 2.5% benzo- phenone-3	10.0	21.5	210	11.4	24.0	211
2% Octinoxate/ 10% Octocrylene	6.3	18.6	295	8.	16.7	190
Average increase			268			228

Table 9.6   Active ingredient isolation from skin: anthralin challenge test.

Subject	Empty SLN	Anthralin SLN	Anthralin + empty SLN	Anthralin alone
1	0 none	0 none	2+ itch/burn	3+ itch/burn
2	0 none	0 none	3+ itch/burn	3+ itch/burn/tender
3	0 none	0 none	3+ burn	3+ slight burn
4	0 none	0 none	3+ burn	3+ tender
5	0 none	0 none	2+ slight burn	2+ slight burn
6	0 none	0 none	3+ burn	3+ burn

Evaluation keys. Visible: 0, no visible reaction; 1+, mild erythema; 2+, well-defined erythema, possible presence of barely perceptible oedema; 3+, erythema and oedema; 4+, erythema and oedema with vessiculation and ulceration. Subjective: subjects were questioned as to itching, burning and tenderness.

MEC process. Anthralin is an organic irritant. It is an extremely potent inhibitor of mitosis but causes erythema and swelling in virtually every exposure. A 0.2% (active basis) emulsion was prepared containing the anthralin SLN. This was then applied to the forearms of six subjects and occlusive patches were applied. The patches were removed after 12 hours and the test sites were rated on erythema and swelling. The test subjects were also questioned regarding the degree of discomfort. Appropriate controls were run: blank SLN (*i.e.* no anthralin) alone, blank SLN plus non-encapsulated anthralin and anthralin alone. The results (Table 9.6) clearly demonstrate that encapsulation adequately contained the drug.

By appropriate choice of emulsifiers, it is also possible to create aqueous nanoparticles in an oil-based emulsion. This technology can be used to encapsulate, for example, water-soluble materials such as L-ascorbic acid (vitamin C). L-Ascorbic acid is considered one of the most important water-soluble antioxidants. It provides photoprotective capabilities by inhibiting UVA and

UVB radiation-induced damage by neutralizing the oxygen free radicals in the skin[40] and it prevents UV immunosuppression.[41] Vitamin C protects the aqueous components of the skin, including tissue and cell fluids.[42] In addition it actively regenerates vitamin E which is the major lipid-phase antioxidant in skin and it protects the fatty components including cell membranes. Vitamin C also stimulates collagen synthesis and has anti-inflammatory properties.[43] As such it has been used in a myriad of cosmetics products to counter the prime signs of ageing—moisture loss, collagen breakdown and free radical damage.

However, the molecule readily degrades in the presence of moisture, oxygen and light. Here, not only does encapsulation provide a convenient formulation vehicle, but also it can enhance the stability of the encapsulated payload (Figure 9.13, which shows data taken at an elevated temperature of 45 °C).

Another unstable cosmetic active is hydroquinone. This is used for the gradual lightening of hyperpigmented skin conditions such as acne spots, freckles, age spots and other unwanted areas of melanin hyperpigmentation that may arise from exposure to the sun, during pregnancy and from the use of oral contraceptives.[44] Because the Asian ideal of beauty for a woman is a light skin tone without spots or appearance, the market for skin lightening formulations continues to grow at a faster rate than in the West: it is estimated that annual sales for Japan alone now exceeds $500 million. As the world population ages, skin problems observed in people everywhere will become more prominent and the cosmetics industry stands ready to supply help.

Hydroquinone arrests the production of melanin through tyrosinase inhibition.[45] However, it rapidly darkens in the presence of moisture and oxygen rendering formulations cosmetically and aesthetically unacceptable.

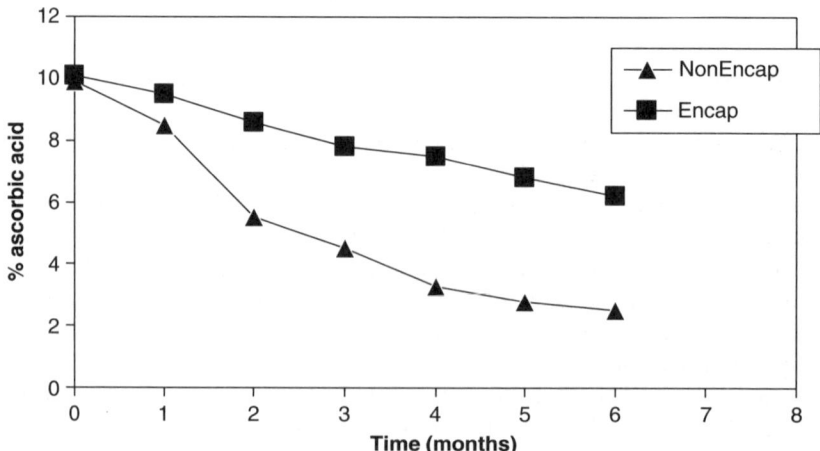

Figure 9.13   Stability at 45 °C of encapsulated vitamin C *vs.* non-encapsulated material. (Note that the assay for ascorbic acid was by HPLC.)

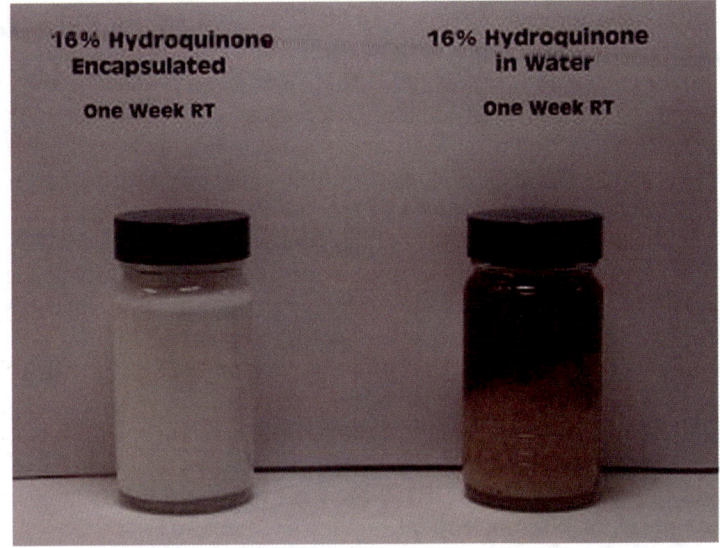

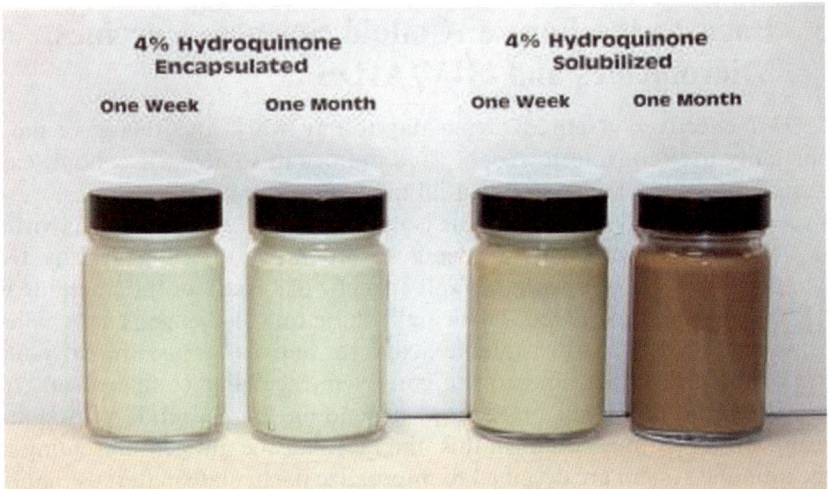

Figure 9.14   Benefit of encapsulating hydroquinone. (a) Colour fastness. Note that there is no settling in the encapsulated suspension. (b) Colour stability in a skin lightening formulation.

Figure 9.14 illustrates the benefit of encapsulation in deterring these unwanted reactions.

Over a period of five years, prior to the sale of sunSmart, we developed, in addition to the above examples, a wide range of custom encapsulated actives for most of the major cosmetic manufacturers. In the majority of cases encapsulation is used to protect the active from the environment (*i.e.* the components of the topical delivery system). Two well-known and highly marketed active examples are the following: Hyaluronic acid, a natural high-viscosity polymer that acts as the "glue" that surrounds, permeates and cushions internal organs.[46] It also helps retain water thus ensuring tissue hydration.[47] Coenzyme Q10 is a vitamin-like substance that is synthesized in all tissues. It is the coenzyme for at least three mitochondrial enzymes and its key role is in producing adenosine triphosphate (ATP), needed for energy production in every cell of the body and upon which all cellular functions depend. Secondary to this, CoQ10 functions as a powerful antioxidant and its free radical quenching properties serve to greatly reduce oxidative damage to tissues;[48] hence its use in moisturizers to help maintain the skin's healthy appearance.

"Hope-in-a-bottle" is a multibillion dollar-a-year business which PSI and I are ever thankful for and more than willing to serve.

Since 2000 PSI has increasingly concentrated on encapsulation of active pharmaceutical ingredients (APIs). Here dispersions of SLNs need to be prepared using biodegradable, all-GRAS (generally recognized as safe) materials to meet the needs of pharmaceutical regulatory demands.

9.3   Back to the Future: Colloid Science, Vaccines, Microbicides and HIV/AIDS

In 2003 I effectively "retired" from day-to-day work and this gave me the opportunity to "return to my roots" and indulge in some basic colloid science but this time with a focus on potential medical applications.

Vaccines produced either from the isolated protein "sub-units" of a virus or gene therapy (the treatment of genetic disorders by inserting healthy DNA within the nucleus of cells with the faulty genetic material), or the emerging field of "RNA interference" ("gene-silencing") share one challenge. That challenge is delivering what is a water-soluble active (or actives) across the fatty membrane of the cell, intact, and to the correct intracellular compartment, from where it can be processed or released depending on the mode of action. In devising strategies to accomplish this, the fact that the antigen-presenting cells of the immune system are designed to internalize particulate materials that have particle sizes of less than 1 μm means that, for example, attaching the sub-unit proteins of a vaccine to such a "nanoparticle" is a logical approach. Further, all such "actives", *e.g.* proteins, DNA, dsRNA and siRNA, share a feature that may allow them to be delivered by a common approach: they are all anionic (negatively charged) in aqueous solution.

So, together with Andrew Loxley (a former graduate student of Brian), who is the manager of special projects at PSI, we prepared some SLNs using biodegradable, non-toxic natural waxes. Through the use of different charge-modifying directors (CMDs) we obtained SLNs having positive, neutral and negative surface charge *in vitro*. After incubating various cell types with some SLNs we found that, irrespective of the surface charge, all SLN types indiscriminately entered the cells with which they were incubated, including

(a)　　　　　　　　　　　　　　　　　　(b)

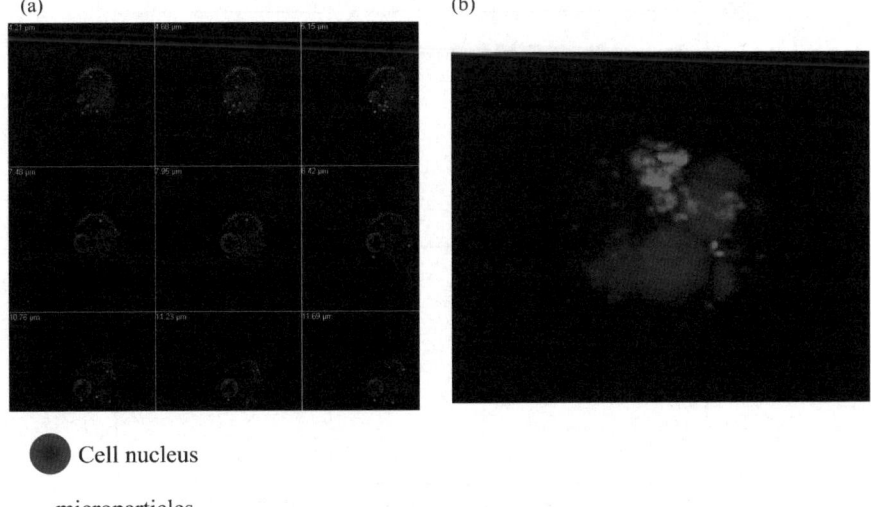

⬤　Cell nucleus

•　microparticles

Figure 9.15　Confocal images of single dendritic cell containing internalized SLNs. (a) Dendritic cell containing green fluorescent SLNs with the nuclei stained blue. (b) THP-1 cell with internalized particles shown as bright green spots within the cytoplasm. Nucleus stained blue.

THP-1 (promonocytes), peripheral blood dendritic cells, Raji (B-cells), PM-1 (T-cells) and endocervical epithelial cells. Figure 9.15 shows confocal laser scanning microscopy images of cells with internalized particles: dendritic cells containing green fluorescent nanoparticles and a THP-1 cell with internalized particles shown as bright green spots within the cytoplasm. Thus, the non-specific nature of the cellular uptake would suggest that such SLNs are suitable candidates for delivering surface-attached actives into cells.

Now, the attachment of negatively charged active molecules to a positively charged surface should decrease the magnitude of the particle surface charge and, with sufficient adsorption, should reverse the sign of the particle charge. This was confirmed, using tetanus toxoid antigen (TTA) as the (negative) active, by following the zeta potential of a cationic SLN as a function of the amount of active attached. TTA was used because it is found in commercial vaccine formulations. The results showed that very high attachment of the TTA can be achieved (up to 0.45 mg of antigen was attached to 0.75 mg of SLN); the zeta potential of the SLN changes from $+64$ mV for the naked particle to -11 mV following adsorption of the TTA (Figure 9.16). This strong attachment, and demonstrated entry into antigen-presenting cells of the immune system (such as the dendritic cells shown in Figure 9.15), illustrates the potential that these natural wax nanoparticles might have in vaccine formulations.

This idea has been taken up as a part of a project to develop a mucosal vaccine for HIV/AIDS. This project, coordinated by Prof. Robin Shattock at St Georges'

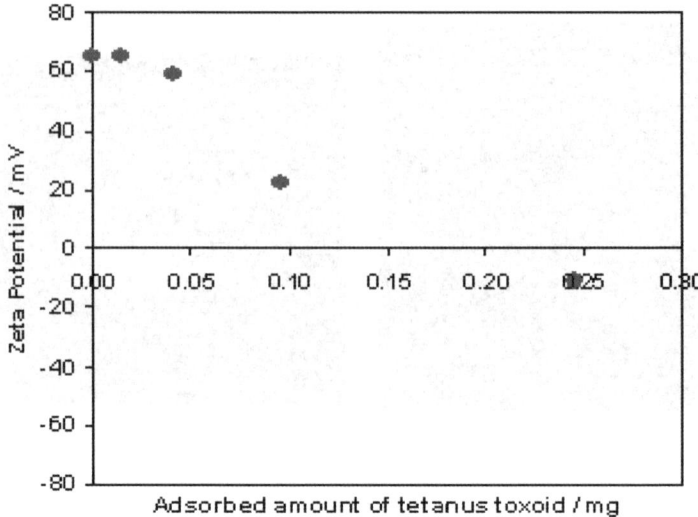

Figure 9.16   Effect of adsorption of tetanus toxoid on the zeta potential of a cationic SLN.

Hospital Medical School, London, is funded as one of the Gates' Grand Challenges in Global Health. Initial results indicate significant increases in *in vitro* immune response (using nanoparticles with the antigen env gp140 attached) over neat antigen. In addition, the serum IgG level in mice was higher in formulations containing antigen/wax nanoparticles compared to free antigen.[49]

The strength of the electrostatic interaction between a negatively charged active and a positively charged SLN may affect the release of the active on arrival at the target cell. We investigated control of the surface charge of nanoparticles made from fatty alcohol waxes by using different ratios (100 : 1, 1000 : 1 and 10 000 : 1) of wax to CMD (Figure 9.17).

The ability to reduce the zeta potential (surface charge) of the nanoparticles could mean that the strength of the binding of the attached active can be controlled, and this could translate to control of the release of active from the nanoparticles once inside the cell—a crucial requirement for DNA, siRNA and dsRNA delivery. This also suggests that the efficacy of a formulation can be tested for its therapeutic effect as the binding strength of active-to-nanoparticle is varied.

I have also become involved in a project to characterize the zeta potential of HIV-1 over a range of conditions relevant for infection, in particular en-compassing those typical of the vaginal environment pre- and post-intercourse. This is part of an ambitious and long-term programme of study, also at St. George's Hospital, initiated by the International Partnership for Microbicides (IPM). IPM is a public–private partnership established by the Gates Foun-dation to develop HIV-prevention options for women worldwide. The aim of our work is to obtain direct experimental evidence of the properties of the

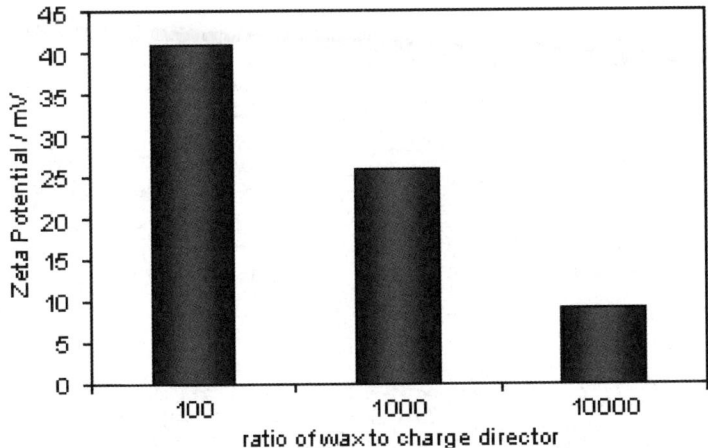

Figure 9.17   Effect of wax-to-CMD ratio on the zeta potential of an SLN.

AIDS virion *in situ* before, during and after interaction with microbicides—a vaginally or rectally applied topical delivery system designed to inactivate incoming HIV-1 or to prevent the virus from entering or replicating in the cells that it infects at or near its site of deposition.[50]

HIV has been extensively studied in terms of its genetic content, biological activity, and structure and reactivity of its surface constituents. The amino acid sequence and tertiary structure of the envelope glycoprotein, gp120, for instance, is well known and is the basis for much of today's HIV vaccine and therapeutic drug development efforts. However, there have been relatively few studies of HIV as a particle with bulk surface characteristics dictated by specific surface moieties. Of equal importance to the surface characteristics of the virion are the surface characteristics of HIV's target cells, the cells that HIV initially infects. Whatever properties exist on HIV, receptive conditions must also exist on the target cells to enable physical contact—a prerequisite to infection. CD4+ T-cells are the primary target cells for HIV.

HIV-1 is extremely difficult to isolate and purify sufficiently for zeta potential measurements. Accordingly, the initial focus of the project was the characterization of the surface charge by non-destructive electrophoretic fingerprinting (Figure 9.18) of three living white blood cell lines that are the principal targets of HIV-1.[51]

Although each of the T-cell lines is zwitterionic, there are real differences in the extent and distribution of the surface (carboxyl and amino) moieties, suggesting serious implications as to the charge characteristics of whole virions that partially derive their surface from host cells. Harvesting from different cell types, different viral phenotypes, different immune cell populations, *etc.*, will all likely have an impact on the viral surface chemistry and thus on any experiments in which they are used.

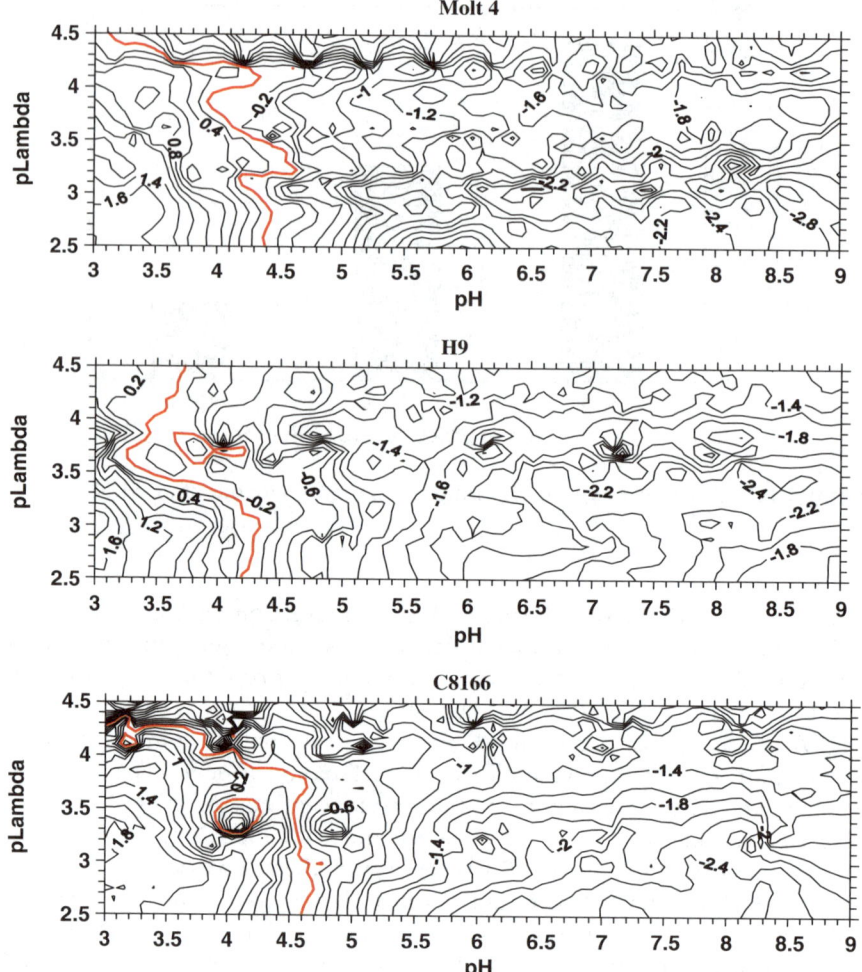

Figure 9.18   Comparison of the electrophoretic fingerprinting contour maps for the three cell lines Molt4, H9 and C8166.

9.4   Concluding Remarks

Colloid science can be used to provide significant benefits in the design and manufacture of personal care, cosmetic, pharmaceutical and medical products and processes. Working on the hypothesis that formulation is not a "black art", nor a "recipe", but rather an exact scientific discipline based on the established physicochemical principles (that underpin colloid science), I have applied this simple approach to the formulation of personal care products such as sunscreens, moisturizers and cosmeceuticals; the use of surface modification, encapsulation and grafting leads to demonstrable improvement in their performance and, as I have shown, it can be financially very rewarding. The same approach also holds real promise in medical applications. And, as we move further into the nano-technology era, the need for innovative/novel instrumentation and techniques to characterize fundamental parameters such as particle size and zeta potential will certainly provide new business opportunities.

I started out measuring zeta potentials of "inert" particles (silver halides) and I am concluding my career measuring those of "living" ones (cells and viruses). In between, I took the "road less travelled" but throughout I have been privileged to have had Brian as a friend and to have been able to call on his vast experience in the field of colloid and interface science. Discussions with him were always lively and instructive; the help and advice he has given me over the years has been invaluable. He is *the* epitome of the truly eminent scientist.

Acknowledgements

I want to thank Brian Vincent for his friendship and advice throughout my professional career, and especially for buying a number of my instruments! My wife and children thank him for opening me to the commercial possibilities of surface modification thus providing the financial security that has accrued.

I acknowledge the invaluable help and support during my instrument days from Dr Peter McFadyen (Brookhaven Instruments UK). More recently, it has been a pleasure to work with Dr Ana Morfesis and Mr Fraser McNeill-Watson (Malvern Instruments).

Finally, a special word of thanks to Dr Andrew Loxley who has helped re-ignite my interest in colloid science and who daily reminds me, by his enthusiasm, that it can be real fun!

References

1. A.L. Smith, F.W. McDowell and D. Fairhurst, *Sonderdruck aus Chemie, Physikalische Chemie und Anwendungstechnik der grenzflächenaktiven Stoffe*, Carl Hanser Verlag, 1973, pp. 679–689.
2. M.C. Wilkinson and D. Fairhurst, *Proc. Int. Symp. on Pore Structures and Properties of Materials*, Prague, September 1973, Part II, pp. C351–C360.
3. M.C. Wilkinson and D. Fairhurst, *J. Colloid Interface Sci.*, 1981, **79**, 272.
4. T. Matsumoto and A. Ochi, *Kobunshi Kagaku*, 1965, **22**(244), 481.
5. D.H. Napper and R.J. Hunter, in *MTI International Review of Science, Physical Chemistry, Surface Chemistry and Colloids*, ed. M. Kerker, Butterworths, 1972, ser. 1, vol. 7.
6. B. Vincent, *Adv. Colloid Interface Sci.*, 1974, **4**, 193.
7. K.E.J. Barrett (ed.), *Dispersion Polymerization in Organic Media*, Wiley, 1975.
8. W. Stöber, A. Fink and E. Bohn, *J. Colloid Interface Sci.*, 1968, **26**, 62.
9. E. Papirer and V.T. Nguyen, *Polym. Lett.*, 1972, **10**, 167.
10. D. Fairhurst, K. Bridger and B. Vincent, *J. Colloid Interface Sci.*, 1979, **68**, 190.

11. Cheol-Kyu Choi and Yang-Bae Kim, *Polym. Bull.*, 2003, **49**, 433.
12. P. McFadyen and D. Fairhurst, *Proc. Br. Ceram. Soc.*, 1993, **51**, 175.
13. I.D. Morrison and C.J. Tarnawskyj, *Langmuir*, 1991, **7**, 2358.
14. J.F. Miller, PhD thesis, University of Bristol, 1990.
15. J.F. Miller, K. Schätzel and B. Vincent, *J. Colloid Interface Sci.*, 1991, **143**, 532.
16. W.W. Tscharnuter, F.W. McNeill-Watson and D. Fairhurst, *ACS Symp. Ser.*, 1996, **693**, 327.
17. M. Kerker, in *The Scattering of Light and other Electromagnetic Radiation*, Academic Press, 1969.
18. G.J. LeVee, L. Oberhelman, T. Anderson, H. Koren and K.D. Cooper, *Photochem. Photobiol.*, 1997, **65**, 622.
19. J.W. Streilein, J.R. Taylor, V. Vincek, I. Kurimoto, T. Shimizu and C. Tie, *Immunol. Today*, 1994, **15**, 174.
20. K. Scharfetter, M. Wlaschek, M. Hogg, K. Bolsen, A. Schothorst and G. Goerz, *Arch. Dermatol. Res.*, 1991, **283**, 506.
21. H.E. Brown, *Zinc Oxide: Properties and Applications*, ILZRO, 1976.
22. M.L. Schlossman (ed.), *The Chemistry and Manufacture of Cosmetics*, Allured, 3rd edn, 2000.
23. D. Fairhurst and M.A. Mitchnick, in *Sunscreens: Development, Evaluation and Regulatory Aspects*, ed. N.J Lowe, N.A. Shaath and M.A. Pathak, Marcel Dekker, New York, 2nd edn, 1997, ch. 17, p. 313.
24. A. Deflandre and G. Lang, *Int. J. Cosmet. Sci.* 1988, **10**, 53.
25. N.J. Lowe and J. Friedlander, in *Photodamage*, ed. B.A. Gilchrest, Blackwell Science, 1995, p. 201.
26. R.M. Sayre and J.C. Dowdy, *Photodermatol. Photoimmunol. Photomed.*, 1998, 38.
27. I.E. Kochevar, in *Photodamage*, ed. B.A. Gilchrest, Blackwell Science, 1995, p. 51.
28. M.A. Mitchnick, D. Fairhurst and S.R. Pinnell, *J. Am. Acad. Derm.*, 1999, **40**, 85.
29. K.J. Rudham and R. Rudham, *J. Chem. Soc., Faraday Trans.*, 1993, **79**, 1867.
30. R.I. Bickley, L.T. Hogg, T. Bonzales-Carreno and L. Palmisano, in *Preparation of Catalysts*, ed. G. Poncelet, Elsevier, 1995.
31. *Federal Register*, 21 May 1999, FDA, Washington, DC.
32. R. Paull, *The Forbes/Wolfe Nanotech Report*, 29 December 2003, Forbes.
33. G.G.J. Hayden, *Lancet*, 1997, **350**, 863.
34. J.O. Funk, S.H. Dromgoole and H.I. Maibach, in *Sunscreens: Development, Evaluation and Regulatory Aspects*, ed. N.J Lowe, N.A. Shaath and M.A. Pathak, Marcel Dekker, New York, 2nd edn, 1997, ch. 34, p. 631.
35. L.C. Harber and D.R. Bickers, *Photosensitivity Diseases: Principles of Diagnosis and Treatment*, B.C. Dekker, 1989.
36. P. Speiser, *EU Pat.*, 0167825, 1990.
37. R.H. Müller and J.S. Lucks, *EU Pat.*, 0605497, 1996.
38. D. Fairhurst and M.A. Mitchnick, *Cosmet. Toiletries*, 1995, **110**, 47.

39. B.L. Diffey, in *Sunscreens: Development, Evaluation and Regulatory Aspects*, ed. N.J Lowe, N.A. Shaath and M.A. Pathak, Marcel Dekker, New York, 2nd edn, 1997, ch. 30, p. 589.
40. Y. Shindo, E. Witt, D. Han and L. Packer, *J. Invest. Dermatol.*, 1994, **102**, 470.
41. J. Murray, D. Darr, J. Reich and S.R. Pinnell, *J. Invest. Dermatol.*, 1991, **96**, 587.
42. D. Darr, S. Dunstan, H. Faust and S.R. Pinnell, *Acta Dermato-Venereologica*, 1996, **76**, 264.
43. H. Freiberger, G. Grove, A. Sivarajah and S.R. Pinnell, *J. Invest. Dermatol.*, 1980, **75**, 425.
44. V.J. Hearing, in *Pigmentation and Pigmentary Disorders*, ed. N. Levine, CRC Press, 1993, p. 4.
45. Ok-Sub Lee and Eun-Joung Kim, in *Cosmeceuticals: Active Skin Treatment*, Allured, 2002, p. 230.
46. H.J. Rogers, in *The Biochemistry of Mucopolysaccharides of Connective Tissue*, ed. F. Clark and J.K. Grant, Cambridge University Press, 1961.
47. F.A. Bettleman, in *Biophysical Properties of the Skin*, ed. H.R. Elden, John Wiley, 1971, vol. 1.
48. L. Ernster, in *Biomedical and Clinical Aspects of Coenzyme Q*, ed. K. Folkers and Y. Yamamura, Elsevier, 1977, p. 15.
49. A. Loxley, D. Fairhurst, C. Eatmon, R. Shattock. M. Arias and F. Wegmann, poster presented at AIDS Vaccine 2007, Seattle, WA, August 2007.
50. J.P. Moore, *N. Engl. J. Med.*, 2005, **352**(3), 298.
51. D. Fairhurst. R.L. Rowell, I.M. Monahan, S. Key, D. Stieh, F. McNeil-Watson, A. Morfesis, M. Mitchnick and R.A. Shattock, *Langmuir*, 2007, **23**, 2680.

Subject Index

Page references to *figures, tables and text boxes* are shown in *italics*.